Co-Manufacturing and New Economic Paradigms

Giulio Focardi
Osun WES, Italy

Lorenza Salati
Osun WES, Italy

A volume in the Advances in Finance, Accounting, and Economics (AFAE) Book Series

IGI Global
DISSEMINATOR OF KNOWLEDGE

Published in the United States of America by
IGI Global
Business Science Reference (an imprint of IGI Global)
701 E. Chocolate Avenue
Hershey PA, USA 17033
Tel: 717-533-8845
Fax: 717-533-8661
E-mail: cust@igi-global.com
Web site: http://www.igi-global.com

Library of Congress Cataloging-in-Publication Data

Names: Focardi, Giulio, 1973- editor. | Salati, Lorenza, 1981- editor.
Title: Co-manufacturing and new economic paradigms / by Giulio Focardi and Lorenza Salati.
Description: Hershey, PA : Business Science Reference, [2019]
Identifiers: LCCN 2018015196| ISBN 9781522570899 (hardcover) | ISBN 9781522570905 (ebook)
Subjects: LCSH: Manufacturing industries--Management.
Classification: LCC HD9720.5 .C62 2019 | DDC 338.8/7--dc23 LC record available at https://lccn.loc.gov/2018015196

This book is published in the IGI Global book series Advances in Finance, Accounting, and Economics (AFAE) (ISSN: 2327-5677; eISSN: 2327-5685)

British Cataloguing in Publication Data
A Cataloguing in Publication record for this book is available from the British Library.

Advances in Finance, Accounting, and Economics (AFAE) Book Series

ISSN:2327-5677
EISSN:2327-5685

Editor-in-Chief: Ahmed Driouchi, Al Akhawayn University, Morocco

MISSION

In our changing economic and business environment, it is important to consider the financial changes occurring internationally as well as within individual organizations and business environments. Understanding these changes as well as the factors that influence them is crucial in preparing for our financial future and ensuring economic sustainability and growth.

The **Advances in Finance, Accounting, and Economics (AFAE)** book series aims to publish comprehensive and informative titles in all areas of economics and economic theory, finance, and accounting to assist in advancing the available knowledge and providing for further research development in these dynamic fields.

COVERAGE

- Economics of Intellectual Property Rights
- Evidence-Based Studies
- Ethics in Accounting and Finance
- Applied economics
- E-finance
- International Trade
- Accounting Standards
- Internet Banking
- Auditing
- Public Finance

IGI Global is currently accepting manuscripts for publication within this series. To submit a proposal for a volume in this series, please contact our Acquisition Editors at Acquisitions@igi-global.com or visit: http://www.igi-global.com/publish/.

The Advances in Finance, Accounting, and Economics (AFAE) Book Series (ISSN 2327-5677) is published by IGI Global, 701 E. Chocolate Avenue, Hershey, PA 17033-1240, USA, www.igi-global.com. This series is composed of titles available for purchase individually; each title is edited to be contextually exclusive from any other title within the series. For pricing and ordering information please visit http://www.igi-global.com/book-series/advances-finance-accounting-economics/73685. Postmaster: Send all address changes to above address.

Titles in this Series

For a list of additional titles in this series, please visit:
https://www.igi-global.com/book-series/advances-finance-accounting-economics/73685

Agricultural Finance and Opportunities for Investment and Expansion
Augustine Odinakachukwu Ejiogu (Imo State University, Nigeria)
Business Science Reference • ©2018 • 236pp • H/C (ISBN: 9781522530596) • US $195.00

Regaining Global Stability After the Financial Crisis
Bruno Sergi (Harvard University, USA) Filip Fidanoski (University of New South Wales, Australia) Magdalena Ziolo (University of Szczecin, Poland) and Vladimir Naumovski (University American College Skopje, Macedonia)
Business Science Reference • ©2018 • 383pp • H/C (ISBN: 9781522540267) • US $215.00

Employment Protection Legislation in Emerging Economies
Samir Amine (Universite du Quebec en Outaouais, Canada)
Business Science Reference • ©2018 • 301pp • H/C (ISBN: 9781522541349) • US $195.00

Fractal Approaches for Modeling Financial Assets and Predicting Crises
Inna Nekrasova (Southern Federal University, Russia) Oxana Karnaukhova (Southern Federal University, Russia) and Bryan Christiansen (PryMarke, LLC, USA)
Business Science Reference • ©2018 • 306pp • H/C (ISBN: 9781522537670) • US $215.00

Regulation and Structure in Economic Virtualization Emerging Research and Opportunities
Denis Ushakov (Suan Sunandha Rajabhat University, Thailand)
Business Science Reference • ©2018 • 238pp • H/C (ISBN: 9781522549666) • US $145.00

Accountancy and the Changing Landscape of Integrated Reporting
Ioana Dragu (Babes-Bolyai University, Romania)
Business Science Reference • ©2018 • 303pp • H/C (ISBN: 9781522536222) • US $195.00

For an entire list of titles in this series, please visit:
https://www.igi-global.com/book-series/advances-finance-accounting-economics/73685

IGI Global
DISSEMINATOR OF KNOWLEDGE

701 East Chocolate Avenue, Hershey, PA 17033, USA
Tel: 717-533-8845 x100 • Fax: 717-533-8661
E-Mail: cust@igi-global.com • www.igi-global.com

Table of Contents

Preface

The book presents the results of a long-term international research carried out in more than 100 shared workspaces, where different economic agents, like professionals, artisans, artists or SMEs have their own operational headquarters and work closely together. These shared workspaces belong to different categories, and to the extent of this book can be named under the general term of "shared workspaces", or "shared workplaces". Different types of shared workspaces will be presented in the book, as a general background, as well as a presentation of some of the visited places, as examples of the different types. The first part of the book will also present the structure, the methods and the evolution of the research from 2012 to 2017.

What the research underlined is a new way to imagine, project and produce products, based on a deep integration within different economic actors driven by concepts of sharing and exchanging. The specific practice underlined in the book was identified as "Comanufacturing", and is characterized by peculiar elements regarding organization, development, strategies.

This practice requires a high level of trust and acquaintance between actors and, although it is also possible between isolated companies, comanufacturing is a common finding within shared workspaces visited during the research.

The axis of the research project was to carry out a systemic analysis of the various components of the system structure of Comanufacturing Spaces (workers, Spaces, Schools, Universities, Educational Bodies, Companies, Institutions), trying to identify the underlying mechanisms that govern relations between this different components.

Visited shared workspaces are independent socio-economic environments, with no relationships among them, which show specific and peculiar traits that are extremely similar, regarding governance, organization, management of internal and external relationships and other aspects. Therefore, the main emerging characteristics and unifying elements among them will be presented in the second part of the book.

Economic agents that are part of shared workspaces are very diverse and people who run them refer to different occupational categories of self-employed persons, artisans, entrepreneurs, professionals, artists, and others. Nevertheless, the people encountered during the research showed very similar traits in terms of self-perception, values, habits. For this reason, in the book they are gathered under the label of crowdworkers, to emphasize their attitude to collective work.

Through a qualitative and quantitative analysis of first hand original data, performed with methods from visual anthropology, authors try to understand the specific reasons why crowdworkers join shared workspaces. There are essentially two different categories: the initial motivation before joining the space and reasons that keeps them in over time. At the beginning simple needs are bringing companies into this new system: the convenience of the space for example, but these are not the reasons why people keep on staying into the space. The most important advantages need to be discovered by experiencing them. The mix of skills, points of view, and experiences, the interaction between technological innovation and traditional craftsmen knowledge, the circulation of ideas are part of the daily life in these physical environments and then, after a while just happen, naturally.

Then, commercial and marketing aspects are analyzed as important elements for common growth opportunities and community building. If, in general terms, sharing a tool or a workspace can be perceived as demanding, sharing monetary incomes is even much more serious. Sharing Economy, in general, is a theoretical platform within is possible to imagine original models and methods for producing and working in a collaborative way. Shared workspaces, flexible and open, give the chance to test them. In these spaces are emerging new tools that help workers to share in specific contests. Is here properly explained one of these new tools developed by the authors: Collecto is a tool for the distribution of profits from common projects not based on working hours but on tasks and mutual understanding of others duties.

It is interesting at this point to focus on why through sharing companies can grow better than in a traditional structure. Once a crowdworker is established inside the space, sharing becomes something natural, but which are the benefits of sharing knowledge?

A new system of course requires new tools and there may be many ways to develop them, but the key point is that among shared workspaces is changing the idea of the primary goal of running, hosting or promoting economic activities: what's motivates the company growth is not to make more profit but make the community bigger and stronger. Shared workspaces become cultural agents of this change of perspective, they can re-inspire citizens to innovate again rather than following set rules. Everything new is a change and change is uncomfortable, can be scaring, but crowdworkers are more open than traditional entrepreneurs to any kind of new ideas and are ready to this change.

Crowdworkers, are diverse in their age, path, experience, and specific profession, but expresses a unified and recognizable model of doing business and life style. Authors believe that social dynamics come before economic ones and this is why physical spaces for working purposes become so crucial: they activate social dynamics that then are related and interconnected with economic ones. For crowdworkers sharing is an attitude, an habit, a way of being. It is a free exchange based on the sharing of a common condition, on recognizing part of the same project and on the will of being an active part in building a common destiny. Comanufacturing spaces are also places where to manage Job Insecurity. They can re-elaborate flexibility in a positive way as it becomes a relevant asset for companies. Crowdworkers are the Social Class of "Stable Flexibility", as being part of a comanufacturing space provides some form of safety, and job insecurity is turned upside down. This flexible condition becomes an opportunity to foster innovation, to have new visions and to build new values. Crowdworkers can be seen a new social class, composed by people who share means of production and have the availability of them as the result of a collective availability.

Therefore, this new social class is strongly linked from the point of view of self-representation to the place where it works, the spaces of comanufacturing, which become symbols of a different way of conducting an entrepreneurial activity, as well as a source of concrete help.

The book is not intended to be a mere analysis of a phenomenon, but to be a reliable point of reference from which to start for further developments.

So, the research carried out is described and what it has brought in terms of knowledge, but also the intervention model that has guided the real development of a shared workspace conformed to the multifactory model, realized without any public funding as an entrepreneurial activity and that in 2018 enters its second year of activity, resulting in economic surplus.

The book therefore alternates first-hand stories with practical guides and theoretical insights, in the conviction that only a multidimensional and multidisciplinary approach to the study of systems can provide a valid key to interpretation.

The model described in the book represents the vanguard of solutions for the creation of shared workspaces, balancing community development and business development. Therefore, the authors believe that books can represent an original voice in the debate on the forms and organizational models of economic systems linked to the development of communities and local chains in urban areas.

The natural audience of this book consists of four categories of people. The first is made up of students and researchers in sociology, political science, economics, architecture and urban planning, or in any case those interested in community processes, systemic thinking, social sustainability, as well as in the themes of sharing economy and community economy, or shared workspaces. The members of this category will appreciate the impressive work of data collection and analysis and the possibility of accessing first-hand sources, collected directly on the field one by one.

The second category is made up of politicians, managers and decision-makers of public structures, interested in better understanding the dynamics and evolutionary paths of community solutions to support entrepreneurship. The representatives of this category will appreciate the participatory governance structure of the model and the fact that it is foreseen from the beginning the achievement of full economic independence and operational autonomy, which allows the project to live a life of its own once started.

The third category is made up of entrepreneurs and owners of disused industrial or craft spaces, looking for a model to be able to revitalize them with a minimum investment. The exponents of this category will appreciate the concrete and direct character of the book, aimed at providing solutions to concrete problems, as well as the real existence of economically self-sufficient workspaces realized in the form of a private initiative on the multifactory model.

Chapter 1 presents the research in its general and method aspects and provides a useful background for reading the book. As for the research, it is explained how it started, the two macrophases in which it was divided, and the six main actions that make it up. As for the background, a selection of different types of shared workspaces is presented.

Chapter 2 focuses on the particular aspects of the shared workspaces studied during the research, presented through extracts from more than 150 interviews with members of the different spaces that were created during the research. The chapter concludes with the presentation of a peculiar organizational form for shared projects that has been identified in different spaces, called the Invisible Factory.

Chapter 3 focuses on the members of the shared spaces, called Crowdworkers in the book", who represent a new transnational social class. The chapter focuses in particular on the elements of self-perception, presented through excerpts from the interviews with the members of the shared work spaces.

Chapter 4 is a conceptual reflection on the role of comanufacturing in the current social and productive system. The chapter is divided into three parts. The first one investigates the relationship between crowdworkers and the capitalist economic system, identifying the differences between the two approaches to conceiving the enterprise. In the second, the approach to sharing of shared workspaces is compared with the sharing economy, arriving at determining how more coherent is the talk of community economy, rather than sharing economy, because of the strong importance that community dynamics have. The third part brings together the themes and provides a theoretical framework on why community dynamics are essential to correctly frame the practices of mutual aid in the legislative landscape of shared workspaces.

Chapter 5 is intended to visualize and through images the different concepts related to the various types of spaces, therefore it presents some real examples of comanufacturing spaces through their presentation documents and their self-perception. An example has been chosen for each main type of space, to give an idea of what differences in approach are hidden behind the different names.

Chapter 6 presents the multifactory model, that is the model elaborated by the Authors to realize from scratch a medium or medium to large sized comanufacturing space. The model is presented in its constitutive aspects, two examples of multifactory built following the model are presented and the subdivision into phases of the intervention that leads to the realization of a multifactory is reported.

Chapter 7 focuses on future research lines, in particular a possible line of study on the distribution of shared workspace with respect to their size class and the proposal of an international network of similar spaces, designed to scale the potential of the system.

The most interesting aspect of the book is that it is the result of empirical research, carried out in the field over a period of five years, visiting over one hundred shared workspaces, in which the authors have spent over a thousand days of stay, making about 150 interviews.

The experience described in the book is absolutely unique. On the one hand, for the extensive and thorough research, carried out over a very long period of time and with a participatory approach that has sometimes required periods of stay in various spaces, even very long. Another aspect of uniqueness is the fact that the experience began in 2012, that is at the dawn of the phenomenon described. This has allowed authors to be the first and only ones to study and describe, starting from the phases before the opening, or in the phases immediately following, some shared work spaces that today are among the most important in Europe. A third element of uniqueness is the creation of a complete model for the constitution, development and strengthening of a space of comanufacturing. From the experience of study, not only an analysis of the existing one was born, but also some elements of weakness in the existing models were identified and a concrete answer was tried to identify, which took the form of the multifactory model. A fourth element of uniqueness is that the model was then concretely applied, first in an experimental form in a small size, and then materialized in a space of 7,500 square meters consisting of six buildings and capable of accommodating about forty companies, entirely designed and implemented following the multifactory model, which has proven to meet the needs for which it was built.

The book is structured in such a way as to leave as much space as possible for the data resulting from empirical research, limiting the parts of theory to the minimum necessary. This has allowed to create a text that can be useful both for those who have a real need to start a shared workspace and need first hand information, and for those who want to have a concrete database, to make comparisons and comparisons with the results of other studies.

Chapter 1
Background, Method, and Research

ABSTRACT

This book talks about shared workspaces, with particular attention to spaces designed to host the production of objects, and is widely based on the results of extensive experimental empirical research conducted by authors in Europe, the USA, and South America between 2012 and 2017. The research involved more than 100 shared workspaces, in which the authors spent over 1000 days. The ethnographic on-field research was carried out using direct observation and through many interviews, which were filmed and subsequently processed through typical methods of visual anthropology. This chapter provides an overview of the different types of shared workspace, illustrates research actions and explains how and where it has developed.

All the world says: yes we know what's written in the books but now let's see what our eyes tell us. Brecht B., Life of Galileo

METHOD AND DEVELOPMENT OF THE RESEARCH

The authors started an independent research on shared workspaces in 2012, from the study of a single case in Milan (Italy), the Made in Ma.Ge. (MAGE). The MAGE was located near Milan, in Sesto San Giovanni (IT), in a 2,000 square meters building owned by the municipality, a former part of the Falk steel

DOI: 10.4018/978-1-5225-7089-9.ch001

factory. The MAGE started in December 2010 as a temporary reuse project, to rethink and experiment with new forms of sustainable entrepreneurship. The MAGE was initially intended to last for three years, but the experiment was extended until the end of 2017.

By the end of 2012 it was made up of 16 activities (Artedì, Arteintasca, Atelassé, Effemeridi, 2be2share, Fabbricanti Di Gioie, GarbageLAB, GHOSTZIP, Laafia, Lascialascia, RiCreazioni, Blanka Studio, SemidiSesamo, Vectorealism, Imana Project), working in different fields like fashion, accessories, jewellery, architecture, art, digital fabrication, web communication, home and office furniture, design, videomaking, bike upcycling, photography. Most of these activities were artisanals, small companies and freelance and involved up to 50 people in total, including employees and collaborators. The authors were part of that environment and since 2011 performed an informal participatory observation of the everyday dynamics between the companies that were part of MAGE. During that observation period the authors could notice that the socio-economic system of these companies had some peculiar points of interest. Collaborations and exchanges were frequent and mainly took place between companies and professionals working in very different fields, resulting in innovative products and unexpected business opportunities.The authors thought that this behavior could be the result of an emerging property of the socio-economic system created by these 16 companies, and decided to study the MAGE as a complex system made of interacting heterogeneous agents, in which each company was seen as an "agent". The work did not want to be a mere representation of the actual reality, but was intended to identify the underlying mechanisms that governed relations between the different components of the system, to understand the reinforcement and feedback mechanisms involved in enhancing or limiting the system.

As the MAGE was a closed system, and people experienced a strong community life, the first idea was that the constant overlapping of personal and professional relationships between people could foster innovative practices in developing new projects and solving issues, so at the beginning the main interest was to understand if there was any relationship between the global behavior of the agents as economic agents and their interactions as social agents.

As the authors were (and were seen by other members of the MAGE) as part of the community, they approached the research assuming an ethnographic point of view, as members fully involved in the daily life of the community, and used tools and methods from visual anthropology to record, track, compare and analyze information coming from members of MAGE.

The authors didn't have a theory to validate, and started the research with just an ordered data collection and the interviews were intended to collect data in a way that could be at the same time structured, open and suitable for further manipulation.

This may seem contrary to the need to conduct an orderly and scientifically structured work. Kirk and Miller (1986), among others, affirm that "is unscientific as well as maddening to initiate data collection without a language (paradigm) that precisely contrasts data and noise. The ethnographer who gathers without knowing what he or she wants (at the logical level) will find no happiness in the process." (p. 66) while de Laine (1997) states that "Qualitative research generally lacks the structure of quantitative research provided by a hypothesis at the outset. The naturalistic researcher or ethnographer typically does not work with either a priori theory or variables. These are expected to emerge from the inquiry. (p. 24)"

Remembers that at the base of this approach there is the idea that "is possible, as well as desirable, to approach the setting with no preformed ideas." And "When this approach is preferred, the research process will, at least in theory, be influenced by it, as the data gathering will begin in a highly unfocused way." Hegelund (2005).

The authors have been aware since the beginning of the impossibility of a completely objective ethnographic research, therefore adopted some strategies that, even if it took innumerable time to be re-elaborated, lead to the collection of data as objective as possible, through the use of audio-visual recordings and programs for video editing.

The authors decided to structure the data collection as a series of long interviews of forty minutes each, without ever interrupting the recording, and to ask a dozen questions regarding the approach to the job, but also personal and professional values, lifestyle, and how work lives were organizes. There were questions that were quite difficult and complex to answer and also open answers, so that those interviewed couldn't just focus on how the specific collaborations took place, but had to give back a true representation of who they were and their overall approach to their professions. This way, it would have been possible to underline the differences in personal approaches, and to intercept which behaviors could be responsible for the constant collaborations and how the system structure could be responsible in fostering and promoting such behaviors.

At the end of the research more than 100 interviews were completed and all of them were performed by the authors themselves, each time reproducing the same frame and context, the recorded interview.

Visual ethnography allows to review over and over what was recorded as in a moviola and therefore allows a further rework. The computer memory of the recorded data is added to the human memory.

The main step was the subdivision of interviews into different clusters. Hegelund (2005) warns about risks involved in data selection: "although this debate is long running and seemingly impossible to settle, it should be clear that to say, innocently, "The data themselves do not lie" is much too simple a way of looking at data. Even if one is not willing to accept that the data are "infected" (the softer version), one should at least realize that the selection of what to count as relevant data (it does not need to be theoretical sampling as such) is, to a large extent, guided by one's theories, tacit knowledge, worldview, and so on. In this way, the data are not really neutral, as they have been selected (from a n infinite amount of potential data) according to one paradigm and not another." To avoid such a risk, selection of topics to become clusters was made on a quantitative base, and only topics cited by at least four people became a cluster, and technically a new timeline in the video editing program. At the end of the work, 40 different timelines were created.

The timelines immediately highlighted the most important topics for the interviewees and at the same time showed a uniformity, in some cases also identical, in the answers given by the interviewees. At the same time, there were virtually identical answers to the same question regardless of social status, country or profession.

Members of MAGE were very diverse, not only in term of their actual professions, but also for what concerns age, past professional and life experiences, political views, but after the first interviews, their answers to the proposed questions were almost identical. It was almost impossible to compare different approaches, as all people interviewed answered to each question in an almost identical way.

The fact that there were constant collaborations between people so different was unusual, but the fact that their values and approaches to life could be so similar was completely unexpected.

The research took form as an on field observation and was extended throughout eight comanufacturing Spaces in five European Countries (Italy, UK, Germany, Portugal, and Spain) lasting for more than two years.

THE FIRST PHASE OF THE ON-FIELD RESEARCH

In this phase of the work, eight projects were taken into account:

- *Made in Mage (MAGE)* - located in Sesto san Giovanni (IT), in a 1,800 square meters former factory
- *Building Bloqs* - located in London (UK), in a 1,000 square meters space in Tottenham.
- *Microfiliera Pollino* - in Basilicata region, South Italy.
- *Balnéario*- located at the first floor of *LX Factory,* the creative district in Lisbon (PT).
- *Oficina Colectiva* – in a former bakery in a central neighborhood in Lisbon (PT).
- Agora Collective – a 3 floors building in the Neukolln neighborhood of Berlin (D).
- FreiLand – a cultural centre located in Potsdam, close to Berlin (D).
- *MOB* (*Makers of Barcelona*) - located in Barcelona (ES), in a 800 square meters space.

All of the visited spaces at that time used to refer to themselves as belonging to several different categories, as Coworking Spaces, Hacker Spaces, Fab Labs, Makerspaces, Creative Spaces, Open Workshops.

In this book, shared workspaces that include, even in a non-exclusive form, the production of objects will be generally gathered under the heading of "comanufacturing spaces"

As said before, the research started in North Italy, at MAGE:

The MAGE was a local value chain, but also a community project with an experimental character, and this aspect could somehow have played a role in bringing people to accord their answers, due to everyday contact with others. The absence of a formalized way to rule professional relationships between members of MAGE could have played a role in leading members of MAGE to establish personal relationships with people they usually wouldn't have considered as professional friends, and to overemphasize similarities with other members. It would have been interesting to ask the same questions to some people belonging to an economic system which could be at the same time comparable, but different.

The three main characteristics of the MAGE were identified to be:

1. Focus on production of physical goods and services related with the production of physical goods.
2. Heterogeneity of the companies which are part of the workspace, in terms of industry, structure, history, dimensions, development stage.
3. Community projects, where companies or autonomous workers who are members can deal with the governance system.

To compare the MAGE with other similar experiences, the authors started looking for other projects, which seemed to have, at least partially, these three characteristics, and they became aware that in South Italy there was the only Italian certified (ISO 22005) officinal herbs supply chain, a local supply chain made by farmers, companies and institutions belonging to few municipalities in the region Basilicata.

This economic system was located in a part of the Basilicata that belongs to the *Pollino National Park*, and geographically spread over a dozen municipalities. This economic system was not a shared workplace, but is a closed system of about 20 production facilities involving about 100 people in a relatively small territory, and the agents are mainly companies in the agricultural market. These agents are linked by strong economic relationships and constitute a local system based on the idea of collaboration, involving constant human interactions and relations, but as far as the supply chain is certified, the norms that rule professional relationships between agents are formally set according to the European norm ISO 22005.

In January 2013, the authors were able to arrange in a trip to Basilicata and meet many of the components involved in the Pollino system, including farmers, institutions, local companies involved in different parts of the chain and a school. They all were asked the same questions as the members of MAGE, in order to compare differences.

Surprisingly, answers given by farmers from *Microfiliera Pollino* didn't differ from answers given by members of MAGE, and answers from the two groups of interviewed people seemed to be coherent and aligned on almost every aspect.

This fact was completely unexpected, but could have many different explanations. In a way, the fact that people so different one from the other gave such similar answers could be the sign of a global change, something that goes beyond individual history and comes from a global change. But these similarities could also just be the result of many similar but specific answers to a punctual, but strong, issue. In these years Italy was experiencing the worst financial crisis after the WWII, and this wasn't just a collective shock, but it was also a period when people usually facing very diverse issues found themselves in similar, temporary situations.

If this was the case, it would have been useless to visit other experiences in Italy. It was time to see if it could be possible to find something similar outside Italy, to compare answers given by people from different countries.

Even before the trip to Basilicata, the authors tried to find other European spaces to compare the governance system of the MAGE, and decided to look for some shared workspace in London, as the UK capital is a global trendsetter for cultural and social movements. In 2012 shared workspaces were quite uncommon in Europe and social networks were just at the beginning of their growth, so it was difficult to find workspaces similar to the MAGE, until the authors could get in contact with *Building Bloqs*, a shared workspace self-described as a "coworking for people in practical professions".

The trip to London took place in February and March 2013, when *Building Bloqs* was in its very early start-up phase. There were few members of the community and the authors could experience a deep and intimate participation to daily life of members and initiators of the project, who answered the same questions as MAGE and to *Microfiliera Pollino*. Once again, the direct observation and what people told in the interviews was surprisingly similar to what was recorded before.

The coherence between answers given by people from Milan, London and a small village in South Italy could reinforce the idea of a global movement, a collective answer to similar issues, but more data were needed.

The authors decided to visit other two countries, one belonging to South Europe and one belonging to Central Europe, in search of other shared workspaces to compare. In May and June 2013 they went to Portugal, to visit *Balnéario* and *Oficina Colectiva*, two shared workspaces in Lisbon, and in July they moved to Germany, first to Berlin to visit *Agora*, and later to Potsdam, to visit *FreiLand*.

Balnéario was a small shared workspace located in Lisbon (PT), in a 300 square meters space part *of LX Factory*, a former industrial complex. *Balnéario* had been initiated by a group of architects, and was run by an association, *Planobao*. At the time of the interviews, *Balnéario* was composed by twenty people, professionals, artisans, artists, and artisans.

Oficina Colectiva was initiated and also run by two architects, and it is located in a 200 square meters space with windows on the street, a former industrial bakery in the creative neighbourhood of Lisbon. *Oficina Colectiva* included 8 different activities (artisans, companies and freelancers) for a total of 16 people.

Agora Collective at the time of interviews was an 800 square meters space on four floors in an historical building in Neukolln, a creative and multicultural neighbourhood in South-East Berlin. *Agora Collective* could host up to 80 people and most of them were freelancers and artists, but there was also a workshop for manual works, a forestry for artistic residencies, an event space and a café. Since 2017 *Agora Collective* moved to former industrial building in Kreutzberg, East Berlin, and the project was reshaped in order to also include a co-living space.

FreiLand is a huge space in Potsdam, 20 kilometres away from Berlin. *FreiLand* covers a 15,000 square meters area formerly owned by the local gas and water company and includes companies, artisans, artists, makers, associations, a FabLab, a café, a club for concerts and events, a guest house, an area for conferences and workshops. At the time of interviews there were more than 40 projects running in FreiLand, involving around 120 people.

The last trip has been taken in Spain to visit MOB, *Makers of Barcelona*. At the end of 2013 it was one of the most innovative shared workspaces in Europe; its communication campaign, from the website to the social networks, was one of the most developed and was promoting a new concept in the business sector: the value of a community of freelancers or small companies, collaboration and exchange as a statement and a "maker attitude" where prototyping and experiments were the tools for innovation. Authors decided to visit it and work from there with their laptops using the flexible desks for few days for experiencing the MOB spirit and for some interviews. MOB today includes a café, a *FabLab* and accommodates up to 120 people. It is still highly focused on promoting Makers, digital fabrication and smart manufacturing.

In all these selected spaces the same questions were asked to members, initiators and coordinators of the spaces and the answers were recorded in the same way: a video interview of around forty minutes.

In each space, the people interviewed always gave similar answers to the questions made, although they were coming from completely different backgrounds.

THE BIRTH OF THE MULTIFACTORY MODEL

During the trip to *Building Bloqs*, a model for building a multifactory took its early form and the following travels to Lisbon, Berlin, and Potsdam allowed to refine it and to define the Intervention Model.

Each of these spaces developed without any contact with each other, not even via the web, and none of the people interviewed (more than one hundred people) had ever heard of any of the other places. Therefore, an astonishing aspect was the consistency between the answers given by the people interviewed. The authors speculated on the fact that they had given equal answers as they all found themselves in a similar condition and found similar answers and, while collecting interviews, the authors started realizing that they could be facing a new social class rising from real and common needs and sharing common peculiar values and lifestyles.

The spaces analyzed had all grown on a trial and error basis, without any reference, and also if this was a facilitating element for research, as managers and communities of the various shared workspaces were always interested in participating in research and in telling their own solutions, on the other hand authors (2015) noted that "a systemization of these experiences and the development of a coherent model should have been very helpful to support the establishment of new ones" in order to allow new initiatives to avoid making mistakes that are already known to others. So, once understood the structure of the system and the dynamics between the various elements that compose it, it was possible to identify specific weak points of each visited space and to identify and test corrective actions.

Once all the elements were put together for the analysis of the system, the authors defined a general model to set up and manage comanufacturing spaces, under an unifying intellectual framework which they called "Multifactory Model", with the aim to define a model which could give guidelines to set up a community based comanufacturing space from scratch.

The Multifactory Model was tested in an experimental environment, a pilot project constituted by the association *Bigmagma* in Milano, which became the practical ground to test the model.

Bigmagma was an association founded in 2012 and formed by entrepreneurs, artisans and professionals. In agreement with the members, since 2014 *Bigmagma* became "*Bigmagma Multifactory*" and was organized and managed following the multifactory model.

The multifactory *Bigmagma* served as an experimental environment, to develop and calibrate the model. Overall, the authors found (2015) that after one year from the start of the activities "the companies part of *Bigmagma* were doing quite well on the market and were self-sufficient. All the activities were self-organized and regulated by few democratic concepts, which were formalized in a collective Manifesto". The *Bigmagma* experience came to an end in 2015, as it was experimental, but former members are still in contact and continue to collaborate.

While observing the daily practices of *Bigmagma* members, the authors understood that some specific topics would require regulatory support and developed tools to address them. For example, tools have been developed to organize mutual aid between members, or tools to decide on the levels of involvement of individuals in terms of risk and economic return within shared projects.

The on-field research went on with a second phase in 2015 and 2016, as Sharing Economy started becoming a trendsetter, and involved other fifteen Comanufacturing Spaces in Germany, Poland, Italy, Croatia, Slovenia, and the USA.

The Multifactory Model served as inspiration for other independent projects of Urban Regeneration like *OZ Officine Zero*, a huge project based on a 30.000 square meters area in Rome (Italy) that refers to itself as "Multifactory". The authors themselves are among the initiators of *R84 Multifactory* in Mantua (Italy), a 8.000 squares meters Multifactory started in 2017 which is born on an area that was part of a former refinery.

The following paragraphs give a general overview on how the research at the base of this book was performed, and on the ongoing research actions.

THE RESEARCH ACTIONS

To summarize, the research was at first divided into 4 main actions, that became 6 during the development of the research. Each Action is intended to have a defined goal and to produce tangible outputs.

1. On field research, phase 1. This action was conducted from autumn 2012 to summer 2013. Authors visited several shared workspaces within Europe to understand how they work and their characteristics. During this action, the authors spent several days in each space, taking part in the everyday life and interviewing coordinators of the spaces and members of the community. These people, although belonging to different personal and professional categories, were all named "Crowdworkers", due to their attitude to collaboration and sharing. While doing the research authors understood that crowdworkers could constitute a new social class, and shaped the idea of the "Multifactory" as a shared workspace with these three characteristics:
    a. They are comanufacturing spaces, which means spaces where there is also production of tangible goods and not just services.
    b. They are run or ruled by a community, with a Bottom-Up Governance System.
    c. They have a huge internal heterogeneity, as in these spaces there are several Companies working together, but independently run, comprising Craftsmen, Freelancers, SMEs, and Artists.
2. Development of the Multifactory Model. As all the visited spaces were growing on a trial and error base, the authors understood that it was important to define a structured way to develop a shared workspace from scratch and to manage it. This Action had the purpose to systemize the collected data, to understand how different spaces were organized and which specific characteristics of each space could be considered fundamental to foster collaborative dynamics between members of the spaces. The output was the Multifactory Model, which describes how a Multifactory should be organized, the profile and the role of the key elements constituting a Multifactory, and the dynamics that take place in it.
    a. First exploitation: conferences and articles
    b. Bigmagma: Testing and development of tools.
3. On field research, phase 2. This action was conducted between 2015 and 2016. The authors visited several spaces in Europe and USA and spent a long period in some of these, up to several months. The aim was to refine the model and to understand the steps made by the visited shared workspaces to establish and grow.
4. Development of the Intervention Model. The Multifactory Model condensed all the experiences regarding governance and management of the spaces collected during the on field research in a coherent way. This

could help to run a space, but before running a space there is the need to build it avoiding mistakes. The Intervention Model was developed to define and describes the steps that can lead to the constitution of a Multifactory. This model should describe the desired characteristics of the agents, strategy, methods, and actions to build up a Multifactory starting from scratch and it was developed on the base of the strategies adopted by the spaces visited during the on field research.
   a. R84: Application of the Intervention Model

5. Development of the International Network of Collaborative Spaces. While studying the spaces in the second phase of the on field research, authors understood that collaborations within members of a shared workspace are not enough to meet the needs of complex projects. The INCS was developed to constitute an environment to scale up the model fostering interactions between members of different workspaces.
6. Development of the Computational Model. This Action will be the last one and is aimed at building Dynamic Simulation Model to forecast the emerging dynamics within agents, like members of different spaces starting a new project, in order to give elements to design a proper strategy to ensure collaborative dynamics to establish.

SAMPLE UNITS

This paragraph gives an overview of the visited places during the first phase of the research and of some of the interviewed companies. Interviews for the first block of the research have been recorded from October 2012 to September 2014 and, specifically:

- Made in Ma.Ge. (Milano, IT) October-December 2012
- Pollino (Basilicata, IT) 22-27 January 2013
- Building Bloqs (London, UK) first trip February-March 2013
- Balnéario and Oficina Colectiva (Lisboa, PT) May-June 2013
- Agora Collective (Berlin, D)
- FreiLand (Potsdam, D) July 2013
- Building Bloqs (London, UK) second trip September 2013
- MOB (Barcelona, ES) September 2013

Some profiles of the surveyed organizations are presented below.

Ghostzip: M. Morosi and I. Cappello

GhostZip is a company that manufactures and distributes the 'bags-zip' that bear its brand name. Backpacks, cases, clutch bags, pouches, make-up and coin purses created with a single zip. All raw materials come from Italian companies and all production takes place at Ma.Ge. They have run a showroom three kilometers from Ma.Ge., in the city center for the entire duration of the project. Mauro Morosi and Isabella Cappello were about thirty-five years old at the date of the interview. In 2016 they closed the shop in Sesto San Giovanni and opened one on the Elba's island during the summer months, where the influx of tourists allows a greater sale. (More information about Ghostzip can be found here: www.ghostzip.it.)

GarbageLab: D. Bravi, S. Colombo and F. Macrì

GarbageLab manufactures bags, backpacks, wallets, key rings and diaries from used PVC billboards and disused seat belts. The average production is 500 bags per month and each bag is unique in having different graphics, which depends on the portion of the poster that is used. This characteristic position them in the middle between the craftsman and an enterprise. Daniela Bravi and Simone Colombo, founding partners of GarbageLab together with Francesco Macrì, who met at the film and television school, were about thirty-five years old when they started up on their own. Currently they run a single-brand shop in the center of the Milano fashion district and have extended their production to leather bags. (More information about GarbageLab can be found here: www.garbagelab.it/.)

Vectorealism: M. Bocola, E. Ricca

Vectorealism is perhaps the most important laser cutting service in Northern Italy, certainly one of the first. Although it was possible to withdraw the products at MA.GE. at certain times, most of Vectorealism's customers were not nearby but were from different parts of the world. Generally the order is handled entirely online: customers load their vectorial files, choose the material to be processed, obtain a quotation for the job and finally start the production. They carry on with their activities, which in the meantime have been extended to a 3D printing and CNC service, in the southern area of Milan. Marco Bocola and Eleonora Ricca, in 2012 at the time of the interview, were just over thirty years old. (More information about Vectorealism can be found here: https://www.vectorealism.com/)

Girocambiando: S. Mancini

He created furniture using recycled wood from fruit and vegetable crates, recovered at the end of the market, a material which would otherwise be sent to the undifferentiated landfill. He made his products using only water-based glues and varnishes, both of which are completely non-toxic. All the processing was done without the use of large machines, but only with basic tools, so the object has imperfections, tangible evidence of a craft process. The founder of Girocambiando, Stefano Mancini was about forty-five years old at the time of the interview, after about two years from the start-up, and 'returned to work in his field of origin, computer science.

E. Cristiani

He organizes, promotes and holds courses in Raku ceramics and paper clay. A past as marketing manager of large Italian companies, he decides to start a new professional career taking a space at Ma.Ge. Sesto San Giovanni (ITA). At the time of the interview he was just over forty years old and currently runs a bigger laboratory of raku ceramics in the centre of Milan. (More information about E. Cristiani can be found here: www.eliocristiani.com.)

NaturalMente

NaturalMente was founded in 2010, is part of the Pollino System and is managed by two partners, one of which is a former truck driver. Because of the crisis that in 2008 affected Italy, they decided to risk and set up their own business in the heart of the Basilicata region, driven by a passion for their own land. So they started cultivating herbs, vegetables and legumes, to create products in oil to sell directly at fairs and shops.

BrunoBIOS

Brunobios is located in the municipality of Rotonda, seat of the Pollino National Park. The farm was founded in 2005, it has organic certification (ICEA Certification Body). Specialized in the production of local varieties of beans at risk of decomposition, certified PDO (Protected Designation of Origin). They produce small quantities and focus on high quality. In addition to the poor bean, the Brunobios farm produces organic honey, berry jams and vegetable preserves. www.brunobios.it

Rcicla: F. Oliveira

Rcicla was born in Balneario, Lisbon. Rcicla is managed and directed by a young engineer, who after a few years of an employed studio work decides to start his own business, trying to combine respect for the environment with his passion for bicycles, the desire to experiment with manual and productive work and the desire to make original and imaginative products. Francisco Oliveira was about thirty years old and was specializing in restoration of old bicycles and accessories for polo bicycles.

Toyno: R. Quinta

Toyno was founded in 2011 and is headquartered in Oficina Colectiva, Lisbon. Toyno was born out of the intuition of two designers active between Lisbon and Berlin, with the aim of proposing shapes, themes and stylistic typical of children's games inside furnishing and design objects with a precise functionality. Rui Quinta was about thirty years old at the time of the interview. http://toyno.com/Studio/

Mme Evasè

Mme Evasè is an eclectic fashion atelier, classic shirts and clothing in general. He creates custom-made garments and costumes for the show, organizes and holds sewing and pattern making courses. Mme Evasè was around fifty years old at the time of the interview.

LOLA: R. Brazao

LOLA is a Barcelona and Lisbon based firm dedicated to contemporary architecture and design issues. It was founded in 2003 by Rute Brazão, Riccardo Cavaciocchi and Sandra Ribeiro (Cavalcanti, 2013). Since the inception the company has developed projects of different types and scales, from objects to interior design, facilities and housing buildings. Working on an international level from its start, LOLA developed projects in places such as Barcelona, Lisbon, Lecce, Riyhad, São Paulo and New. The work that was developed relies on a collaborative design method, engaging clients, colleagues and specialist consultants in the creative process.

SECOND PHASE OF THE ON FIELD RESEARCH

The same questions asked during the first phase (2012-2014) of the research have been made also to other spaces along the next four years (2015-2018). The following is a list of creative spaces visited by the authors divided by country and town. Wherever possible, it is added to the name of the space also the website and the pay off appearing in 2018 website version.

Some of these spaces, although at a first view it seems they have common features, when visited did not fit with the new economic paradigm investigated along this book. At a first glance, through an on line research and an email exchange with the space initiators or managers, it wasn't possible to understand which model these spaces were following, and this became absolutely clear only after a first visit to the space.

It has to be said that, although some of the spaces visited were not very useful as a source of inspiration for innovative practices or methods, having visited them in person has been of great importance to understand the differences between different types of shared workspaces and to understand which models produced undesirable effects, at least with respect to the aims behind the development of the multifactory model.

Germany

Berlin

- Beta House Berlin – Coworking Space, https://www.betahaus.com/
- Ex Rotaprint http://www.exrotaprint.de/
- C-Base https://c-base.org/
- Happy Lab Berlin- open Workshop, http://www.happylab.at/en_ber/home
- FabLab Berlin, https://fablab.berlin/en/
- Agora Collective, http://agoracollective.org/

- ZKU - Production site, artist/research-residency, program platform. http://www.zku-berlin.org/
- Ufer Studios - http://www.uferstudios.com/

Potsdam

- Rechenzentrum, http://rz-potsdam.de/cms/
- freiLand, https://www.freiland-potsdam.de/
- Altefeinbaeckerei – Coworking offices, http://www.altefeinbaeckerei.de/Coworking.html

Dresden

- Neon Work – Creative Business Center, http://www.neonworx.de/
- #rosenwerk, Open Workshop, https://konglomerat.org/
- c3d2 – Hackerspace, https://c3d2.de/space.html

Leipzig

- Honoray Hotel, http://www.honoraryhotel.net/
- makerspace Leipzig – Open Workshop for everybody, http://makerspace-leipzig.de/
- tapeten werk –Creative Production Space, http://www.tapetenwerk.de/
- Spin Lab - Accelerator for Start Up, https://www.spinlab.co/
- Die Spinnerei, http://www.spinnerei.de/

Chemnitz

- Fablab Chemnitz, https://fablabchemnitz.de/
- Buildung Werkstatt, http://www.bildungs-werkstatt.de/

Munich

- Haus Der Eigenarbeit, http://www.hei-muenchen.de/
- Werkbox 3, Open Workshop, http://www.werkbox3.de/
- Mumalab, Munich Maker Lab, https://munichmakerlab.de/
- fablab Muenchen, www.fablab-muenchen.de
- Hinterher, Munich bike trailer, https://www.hinterher.com/

Halle

- Eigenbau kombinat - Maker Space and Hacker Space, https://eigenbaukombinat.de/

Augsburg

- Werkraum Augsburg – Open Workshop, werkraum-augsburg.de/
- Open Lab – HackerSpace, https://openlab-augsburg.de/

Eberswalde

- Eberwerk – Open Workshop, http://hebewerk-eberswalde.de/

Nurnberg

- Quellkollectiv, http://quellkollektiv.net/
- Fablab Nurnberg,

Hamburg

- Attraktor – Hackerspace, https://wiki.hackerspaces.org/Attraktor
- Honig Fabrik, https://jim.honigfabrik.de

Poland

Stettin

- OFF Marina http://off-marina.pl/
- ESCapada

Krakow

- Nowa Huta Cultural Centre, https://nck.krakow.pl/

Prague

- brmlab - hackerspace, https://brmlab.cz/

- Do it Yourself, http://www.sdilna.cz/
- Pragovka Art District, http://www.pragovka.com/

Slovenia

Ljubljana

- LJUDMILA - Hakerspace Digital Media Lab, https://wiki.ljudmila.org
- Poligon - Kreativni Center, https://www.poligon.si/

Croazia

Zagreb

- HUB 385 - Coworking, https://hub385.com/
- Radiona - Makerspace, http://radiona.org/

Spain

Barcelona

- MOB – Makers of Barcelona, www.mob-barcelona.com
- Hangar - Cultural Center, https://hangar.org/

Portugal

Lisbon

- LX Factory, Art Center, www.lxfactory.com/
- Oficina Colectiva, https://oficinacolectiva.wordpress.com/

Porto

- Critical Concrete, https://criticalconcrete.com/
- OpoLab – Multidisciplinary Center, Coworking and Fablab, www.opolab.com/

USA

New York

- Gowanus Studio Space – A Coperative Workspace, http://gowanusstudio.org/
- NY Resistors Hackerspace, We Learn, Share and Make Things, https://www.nycresistor.com/
- Brooklin Navy Yard – Industrial Park, https://brooklynnavyyard.org/
- New Lab, A community for entrepreneurs working in advanced technologies, https://newlab.com/

Baltimore

- Station Nord Tool Library - It's like a book library but for tools, www.stationnorthtoollibrary.org
- Red Emma's - Worker cooperative radical book store, https://www.redemmas.org/
- Open Works, Everyone is a Maker, http://www.openworksbmore.com/

The Nederlands

Amsterdam

- NDSM - a 24/7 urban district with residential and (creative) business, http://www.ndsm.nl/en/

Italy

Cremona

- CO Box- An Innovation Space, https://www.co-box.it/

Milan

- Wemake - Milan's Makerspace, http://wemake.cc/
- Coworking Login - Coworking Belongs to Everyone, https://www.coworkinglogin.it/

- Talent Garden Milano – A Campus for Innovators, https://milano-calabiana.talentgarden.org/
- Impact Hub – Come for Coworking, Stay for Community, https://milan.impacthub.net/
- Base Milano - BASE is a project of innovation and cultural contamination between Art, Creativity, Enterprise, Technology and Welfare, http://base.milano.it/
- Cowo lambrate, Brings Coworking to your Office since 2008, http://www.coworkingproject.com/coworking-network/milanolambrate/
- Piano C – Coworking, Cobaby e Community http://www.pianoc.it/
- Copernico, http://www.copernicomilano.it/
- Thefablab – Make it Real!, www.thefablab.it
- FabriQ – Incubator of Social Innovation, http://www.fabriq.eu/
- Rimaflow – Factory restored, https://rimaflow.it/
- Lascialascia, Temporary Architecture and Coworking, https://www.lascialascia.com/
- fablab vectorealism

Sesto San Giovanni

- Via mazzini 49, http://www.lucatongiorgi.com/

Biella

- Sella Lab – A Corporate Innovation Platform, https://sellalab.com/
- Coworking Fondazione Pistoletto, http://www.cittadellarte.it

Alessandria

- LAB 121 coworking and professional collaborations, http://www.lab121.org/

Fiorano Modenese

- Casa Corsini - Coworking and Fablab http://www.casacorsini.mo.it/

Roma

- Officine Zero Multifactory, http://ozofficinezero.org/
- Famo Cose Makerspace, http://www.famocose.it/

Padova

- La Mente Comune, www.lamentecomune.it/

Palermo

- Cantieri Culturali alla Zisa

Lecce

- Manifatture Knos - A space for sharing, manifattureknos.org/

Brindisi

- Ex Fadda – A Urban Laboratory, http://www.exfadda.it/

Ferrara

- Spazio Grisù - Consortium of Creative Enterprises, https://www.factorygrisu.it/

Basilicata

- Microfilierapollinohttp://www.evraitalia.it/filiera-delle-piante-officinali/

Alzano

- Spazio Fase, http://spaziofase.com

Saronno

- Rent Office – Smart offices and coworking, http://www.rent-office.it/

Varese

- FaberLab - Digital contents, http://www.faberlab.org/

Trieste

- Parco Culturale di San Giovanni http://www.parcodisangiovanni.it/

Mantova

- Creative Lab – Center for young and creativity https://www.santagnese10.it/creativelab
- Sant'agnese 10 – Officina Creativa, https://www.santagnese10.it
- FabLab Mantova, http://www.fablabmantova.org/

Ecuador

Quito

- La Tejedora – Districto Creativo, https://www.facebook.com/latejedoradistritocreativo/about
- ImpaQTo – La Mejor Comunidad de Coworking de l'Ecuador www.impaqto.net/

Authors spent more or less 300 days in those workspaces, observing and making interviews.

These spaces range from 200 to 200,000,000 square meters, and a couple of them cover even a wider area, as they are entire neighborhoods or rural areas. obviously there are big differences in the management and coordination of such different areas, even if the profile of the people working in them remains unchanged. In fact, larger areas tend to host more companies, but not necessarily larger ones. So, by now, the shared workspaces remain dedicated to companies ranging from freelancers to SMEs.

From their own experience, the authors have divided the shared workspaces into 5 different dimensional classes, which correspond to different levels of organizational complexity and economic effort:

Class 1, small spaces. These spaces are approximately up to 500 square meters and can accommodate some professionals and 2 or 3 small craft activities, for a total of 6/10 members. These spaces can be created and developed on the basis of private local initiatives of individuals or small groups. The budget required to start them is reduced, usually between 5,000 and 10,000 Euros, and many of the renovation or start-up work can be carried out directly by members. The budget to manage them is similar to a small shop, below 20,000 euro/year. Relations between members are essentially personal and space management is possible even in the absence of a developed organizational structure.

Class 2, small to medium spaces. These spaces are approximately between 600 and 2,000 square meters, and can accommodate up to 15/20 activities, including craftsmen and micro businesses with employees. These spaces can still be created and developed on the basis of impromptu local private initiatives, but more often they are the result of structured projects, carried out with the support of local institutions. The budget required to start them starts to be important, but less than 50,000 euros, and many renovation or start-up works cannot be carried out by members. Annual management expenditure remains below €100,000 per year, but still requires a precise sustainability plan. Relations between members are still essentially personal, but space management requires a developed organizational structure with a system of coordination and division of tasks.

Class 3, medium to large spaces. These spaces can be up to 20,000 square meters, which can accommodate up to 40/50 structured activities, such as shops, bars, craftsmen and micro businesses with employees, plus a significant number of freelancers. They are often composed of areas with several buildings. These spaces are always the result of structured projects, carried out by public institutions or private companies, but with the support of local institutions and included in urban regeneration and redevelopment projects. The budget required to start them goes from 100,000 euros to increase. Annual management costs remain below €500,000 per year, but professional management of the space and adequate financial coverage are essential. Relations between members are still personal, but for some of them it can only be a very superficial relationship. Space management requires a professional organizational structure, with a multi-level coordination system and the presence of employees.

Class 4, large spaces. These are spaces approximately between 20,000 square meters and 100,000 square meters, which can accommodate up to 50 to a couple of hundred structured activities, such as shops, bars, craftsmen and businesses with employees. Among these businesses there is usually a coworking space, which welcomes freelancers. These are all areas made up of several buildings. These spaces are always the result of structured projects, carried out by public institutions or large private companies and included within complex urban regeneration and redevelopment projects. The budget required to start them exceeds 1,000,000 euros and annual management costs are easily over 2,000,000 euros per year, making professional space management and the involvement of banks and other financial institutions essential to provide economic coverage. Relations between members can be largely personal, but at the same time many people may not even know each other. Space management requires a complex organizational structure and a legal entity created for this purpose. The coordination system is always formal with a defined organization chart and a precise control structure.

Class 5, huge spaces. These spaces are approximately 100,000,000 sqm and up, and can also accommodate hundreds of structured activities, such as shops, bars, artisans and micro businesses with employees, plus a substantial number of freelancers. They are often composed of areas with several buildings. These spaces are always the result of very demanding projects, carried out jointly by public institutions and large private companies within urban regeneration projects at the city level. The budget required to launch them is several million euros. Annual management costs are those of a medium-sized company and management is often carried out by an external group. Relations between members are still personal within subgroups, but it is impossible to know all members. Space management requires a very complex organizational structure, organized into divisions and involving several employees.

Although more than half of the spaces visited belong to class 1, the authors noted that the most interesting spaces as starting environments for collaborative processes are class 2 and class 3. The reason is probably to be ascribed to two factors, the first being that they are large enough spaces to accommodate many companies, multiplying the possibilities of exchange, the second being that at the same time they are small enough to allow members to participate actively and personally in the organization and management of the community.

A SHORT PRESENTATION OF DIFFERENT KIND OF SHARED WORKSPACES

This paragraph presents different types of Shared Workspaces and summarizes the evolutionary path of the idea at the base of Shared Workspaces. Shared Workspaces have existed since much prior than usually estimated, but only over the last few years they have become a common presence in urban environments.

Talking about Shared Workspaces means getting into a fairly unexplored territory.

Although some experiences date back to more than thirty years ago, only in recent years Shared Workspaces have become a widespread phenomenon.

Therefore, although there are categories that identify different types of spaces, in practice there are many overlaps and boundaries are not well-defined. It's curious that some of the spaces visited during the research were self-defined as including themselves in a certain category and, years later, in their communication they changed their category of reference. For example, Building Bloqs in 2014 defined itself as a "Coworking for practical professions" and in 2018 defined itself as a "Maker Space", while on the contrary MOB (Makers of Barcelona) in 2015 placed a strong emphasis on the "Making" aspects, while in 2018 it presents itself with a more typical "Coworking Space" profile.

A common point, above any difference, is that shared workspaces are physical space where there are people willing to share something.

As Sapelli (1994) states, "the relationship between person and organization is not only important for anthropological and phenomenological analysis, but also because it allows to build explanatory maps of business behaviors, avoiding differentiation between system and organized economic actor". "The relationship between person and organization leads to the study of Governance, rules and procedures, their formalization and informality, to understand both the traditions of enterprise and the tensions to change". Sharing Economy is a cultural and organizational change, driven by people who are giving structure to this change, and to better understand who these people are and how they are exploiting the change, a starting point is to analyze the places where this change is taking place.

While urban de-industrialization is leaving out empty industrial spaces to be rethought, these urban voids are increasingly taken by heterogeneous groups of self-employed workers across Europe and USA, who are re-inventing them as Shared Workspaces. These workers are the pioneers and the creators of a profound change in the socio-economic structure that can challenge the classic way of thinking about the world of work and business.

This has led to a sharp increase in coworking and co-manufacturing spaces starting from the second decade of the 2000s, fueling the widespread conviction that the idea of sharing a workplace is very recent, but as a matter of fact the idea of having collective spaces where working together goes far back in time, albeit in a different form. For example, just to name a few well-known examples, Amsterdam's NDSM is one of Europe's largest and most well-known Shared Workspace and occupies the huge warehouses of the Nederlandsche Dok en Scheepsbouw Maatschappij where ships of the Dutch merchant fleet were built up to 1984. The space was abandoned for years and then occupied by a squat in 2001, becoming a point of reference for artists, creative people and craftsmen. It was then legalized and has come to host dozens of companies operating in disparate fields. C-Base was founded in 1995 and is the most well-known Hackerspace in Berlin and one of the most renowned among hackers, programmers and digital activists all over the world. Hei, Haus der Eigenarbeit, has been active in Munich since 1987 and since the foundation it has been proposed as a place where anyone could use tools and equipment for various manual activities, attending courses and learning the use of machines, but mainly for hobby and spare time use.

But all of these shared spaces were isolated cases, pioneers non-following any written path or existing model.

In the last decade, some definitions emerged which come from different backgrounds and experiences, like Coworking Spaces, Hacker Spaces, Fab Labs, Makerspaces, Creative Spaces, Open Workshops, and many others, but at the base they are all environments which focus on the idea of sharing and exchanging different kinds of values within a community. Some of these are usually intended to be used in one's spare time, some of these are more focused on people who work to their projects on a daily basis, like Coworking Spaces, mainly aided at hosting Freelancers, professionals, solopreneurs, work-at-home independent workers.

Coworking Spaces

The date of birth of coworking spaces is set at 2005, August 9 when the programmer Brad Neuberg open the first "Coworking Space" in Los Angeles (Foertsch & Cagnol, 2013). Neuberg's goal was to create a workplace more professional than his own kitchen, but less impersonal than the traditional Business Centers, and focused on the quality of human relationships between members. Coworking, it is clear, is the word that became more popular and got more used. Standing to the statistics of the *2017 Global Coworking Survey* published by Deskmag, an online magazine about coworkings, in 2017 the number of coworking spaces was about 13,800 and the number of members was 1,180,000 worldwide.

Coworking is a form of shared space typically open to freelance professionals, creators, web designers and workers in general who need a computer and an internet connection in their professional daily life. They are typically free-lance workers and usually they could work at home, but join a Coworking to overcome the sense of isolation and to get in touch with other freelancers belonging to the same professional field. Coworking spaces are usually equipped with desks and offer basic services as internet, printer, meeting rooms, presentation rooms with projectors and are run by a company that manages top down the organization and rents spaces on a weekly, monthly or semi-annual basis. A Coworking space does not make any kind of production and is usually located centrally in the heart of the city.

Then, over the last few years, other concepts of Shared Workspaces followed, like those intended for people in practical professions as the Makerspaces.

Makerspaces

The term "Makerspace" was used for the first time in 2005 by the MAKE Magazine. In 2011 MAKE Magazine registered makerspace.com and the term began to spread. There are no formal requirements to mark a space as a "Makerspace", so there are some Makerspaces that are ideally linked to the hacker movement and knowledge sharing, especially in free software, open source, copy left, share alike, and an important point is the carrying out of workshops and workshops for students. Other Makerspaces are instead based on a model where a managing entity of the space makes an initial investment on machinery and tools, to offer a service to members who pay to access the space and to use the available machines, as users.

FabLab

A kind of Shared Workspaces close to makerspaces are Fablabs. Fablabs are specifically dedicated to digital fabrication and rapid prototyping, and the collaborative aspect is closely related to the sharing of equipments such as 3D printers, CNC machines, laser cutting machines and the shared development of related knowledge. The idea of Fablabs started as a collaboration between the Grassroots Invention Group and the Center for Bits and Atoms at the Media Lab in the Massachusetts Institute of Technology in 2001, when they proposed the idea of a “Fab Lab”. While “Fablab” is a generic term, “Fab Lab” is a name reserved only to the spaces that participate in the network run by the Fab Foundation and follow specific requirements for space and tools and behavior.

Hackerspaces

Another kind of Shared Space are Hackerspaces. Hackerspaces are community-operated physical places, where people share their interest in tinkering with technology, meet and work on their projects, and learn from each other. Hackerspaces are places where to experiment with computer programming and usually highly concerned about freedom of speech and mass control in the era of computer based technologies. Their history dates back to the Eighties and refers both to US experiences as the famous Noisebridge hackerspace in Los Angeles and his charismatic funder Mitch Altman and to East and West Germany with the experience of Chaos Computer Clubs (CCC). As stated in *Hackerspaces the Beginning* (2011) “Decentralization and a lack of hierarchy might be one of the strongest forces that make the club withstand every tendency to dissolve and disintegrate. Instead, after almost thirty years of existence, the CCC is still a thriving and growing community of individuals that regularly manages to reinvent itself.” Years went by and the CCC had to survive a couple of crisis. In Berlin people from east and west quickly met after the Wall came down. A computer conference was set up even before the two Germanys united in late 1990 and the base for a new CCC was laid out: the CCC Berlin. In December 2016, the authors attended the 33rd Chaos Communication Congress (33C3) in Hamburg: ten thousand tickets went sold out six months before the conference and there were more than one hundred lectures and workshops from activists from all over the world.

Multifactory

Another kind of Shared Workspace is the Multifactory, a kind of comanufacturing environment dedicated both to freelancers and people in practical professions. The term "Multifactory" was first proposed by Focardi & Salati in 2013 at WES Conference held in Warwick by the *British Sociological Association*, to identify with a common name the spaces visited during the first phase of the research and that feature distinctive traits:

- They were comanufacturing spaces, which means spaces where there is also production of tangible goods and not just services.
- They were run or ruled by a community, with a Bottom-Up Governance System.
- They had a huge internal heterogeneity, as in these spaces there are several Companies working together, but independently run, which can be Craftsmen, Freelancers, SMEs, and Artists.

While at the beginning the term "Multifactory" was just a way to identify places with common characteristics, but which could also be part of other categories of Shared Workspaces, since 2014 it started to identify a specific kind of Shared Workspaces, when Bigmagma Multifactory started in Milan. A Multifactory still has the three characteristics mentioned above (focus on production, community project, heterogeneity), and as far as 2018 there are several shared spaces that refer to themselves as Multifactories (the biggest ones are OZ Officine Zero in Rome and R84 Multifactory in Mantova).

Incubators

Another well known kind of Shared Workspaces are Incubators, environments in which start-up companies work in parallel for a defined time, until they are sufficiently solid to exit the space. These spaces do not foresee an exchange between parallel companies, or a common strategy, but they often organize common events to meet specific needs, like meetings with investors or mentors.

Hybridization and Future Perspective of Different Types of Shared Spaces

It is also important to emphasize that there are other Shared Spaces which weren't originally meant to be workplaces, but that became Shared Workplaces following the evolution of the needs of their members. This often happens to Cultural Spaces, Creative Spaces and other places of artistic and cultural production that promote the idea of art and culture as a common good and experiment horizontal models for the management of common goods, and at one point face the need for helping their members in facing the market.

It should however be stressed that the various forms of Shared Workspaces are increasingly hybridizing with each other, and this often happens as the result of the willingness to meet the professional needs of their members. For example, professionals from a Coworking Space can start having the need to prototype or produce items, and this leads to the space buying a 3D printer, or to build a wood and metal workshop, turning more and more into a Co-manufacturing space.

There is an ongoing debate in Europe, which can take very different routes. For example, at VOW-Treffen 2014, organized in Potsdam (D) by the Verbund Offener Werkstätten, delegates debated the form and meaning that German Fablabs will take. Among the ideas, Fablab without machinery, which could work in partnership with external services, leaving to the Fablab a unifying role for different professional environments. On the same occasion, Matthias Röder brought the experience of Fablab Dreseden, a "moving Fablab" that operated for two years without a stable venue and equipped with the limited equipment. On the same days, in Italy, the Make in Italy Foundation organized a similar meeting at Fab Lab in Turin, during which Massimo Menichelli (2014) on behalf of the Fab foundation, reiterated the need, in order to be a Fab Lab, to install at least some specific machinery.

On the other side, it has been said that there is an Italian way to perceive the fablab concept in a completely different way from the North European countries, mostly Germany and The Nederlands. As stated by Manzo and Ramella (2015): "Italian labs have been able to orientate themselves towards local productive activities, developing partnerships with businesses and creating a series of services and activities with ties to the productive vocations of the local area"

On the contrary, the fablab Muenchen, one of the biggest in terms of space and members visited by the authors, cannot host members interested in developing their professional activity inside the fablab.

Thus, the fact that similar institutions working in the same field can have totally different approaches to address the same issue reveals how fluid the international scene of shared workplaces really is.

In any case, the real change beginning from the second half of the 2000s is that sharing spaces have abandoned their characterization as niche proposals and have begun to take shape as real enabling systems for the birth and development of new forms of business, based on horizontal relationships.

The authors suggested (2013) that although Coworking Spaces are the most common and widespread kind of Shared Workspaces, the business model that is behind the idea of Coworking Spaces is weak. First, their business model requires a huge amount of resources to take care of the space and to foster collaboration between members through top-down activities, and this pushes monthly fees up, while at the same time compressing profit for the entrepreneurs who run the space. The results of the 2017 Global Coworking Survey performed by Deskmag (2017) shows that just 24% of the owners of a Coworking space rate compensation they receive from that activity from "Very Good" to "Rather Good", while 75% of them rate it from "Somewhat" to "Very Bad". At the same time, 23% of Coworking Spaces declare to be unprofitable, and 29% of them claim they don't need to be profitable on their own in order to run the entire business successfully.

Then, as Coworking Spaces are interesting for specific categories of workers, like solopreneurs or nomad workers, opportunities to share and collaborate may quickly come to an end, and as co-workers are not linked each other and to the space by structural reasons, like the need to share expensive equipment, this can bring to a loosening of relationships among co-workers. While in the 2016 edition of the Global Coworking Survey, 66% of co-workers interviewed indicated that almost all or many members knew each other's first names, in 2017 this percentage drops down to 56%.

Comanufacturing Spaces are in a completely different situation. Comanufacturing Spaces host economical subjects that build physical objects, or that are related to such kind of companies, and this changes the prospective.

Different companies which work together in a single Comanufacturing Space can share not only experience and skills, but also equipment and investments. Comanufacturing spaces, if well shaped, can change the way goods are invented, developed and produced, and are an opportunity for Artisans and SMEs to reinvent the way they stay on the market.

They can be less dependent on banks, as they can use other's equipments instead of buying them, and also be more relevant on the market, as they can develop joined projects putting together different skills and means of production.

This continuous availability of skills and equipment, and the habit to work on joint projects, modifies the way companies operate, organize, and develop new solutions and fosters long term relationships between members of a space and helps to build structural relationships among them.

CONCLUSION

While sharing and exchanging are becoming for some companies' real assets on which to build their competitiveness, the need for Shared Workspaces is increasing. The boost in their popularity started since the second half of the 2000s, when shared workspaces abandoned their characterization as niche proposals and have begun to take shape as real enabling systems for the birth and development of new forms of business, based on horizontal relationships.

The most well-known kind of shared workspaces are coworking spaces, but these ones don't fit the needs of all the categories of independent workers interested in sharing and exchanging as valuable assets to design future of their companies, with special regard to people in practical professions. Co-manufacturing spaces could be a more valuable option for these companies, but there are many reference models, the boundaries are not always clear and are often misapplied. During the research a general lack of reference models for specific needs appeared and from this observation the authors took the decision to develop a model to start a particular type of shared workspace, the multifactory.

The research work was therefore to recognize and describe a natural and spontaneous process through which a new generation of entrepreneurs self-organize, define the evolutionary phases of this process and identify systems, practices, tools to promote a homogeneous development of the process.

REFERENCES

Barbara. (2014, November 3). Verbund Treffen in Potsdam [Verbund meeting in Potsdam] [blog post]. Retrieved from http://www.offene-werkstaetten.org/post/verbundtreffen-2014-in-potsdam

Cavalcanti, G. (2013, May 22). The difference Between Hackerspaces, Makerspaces, Techshops and Fablabs [blog post]. Retrieved from https://makezine.com/2013/05/22/the-difference-between-hackerspaces-makerspaces-techshops-and-fablabs/

De Laine, M. (1997). *Ethnography: Theory and applications in health research.* Sydney, Australia: MacLennan & Petty.

Deskmag, C. E. (2017). *2017 Global Coworking Survey*. Retrieved from http://www.deskmag.com/en/background-of-the-2017-global-coworking-survey

FabLabs. (n.d.). Retrieved from http://www.fabfoundation.org/index.php/about-fab-foundation/index.html

Focardi, G., & Salati, L. (2013, September). *MultiFactory Model: a New Approach to Job Creation*. Paper presented at the WES 2013 (Work, Employment and Society Conference), Warwick, UK.

Focardi, G., & Salati, L. (2015). A New Approach to Knowledge Sharing, the Multifactory Model. In O. Teràn & J. Aguilar (Eds.), *Social Benefits of Freely Accessible Technologies and Knowledge Resources*. Hershey, PA: IGI Global. doi:10.4018/978-1-4666-8336-5.ch009

Foertsch, C., & Cagnol, R. (2013, September 2). The history Of Coworking In A Timeline [blog post]. Retrieved from http://www.deskmag.com/en/the-history-of-coworking-spaces-in-a-timeline

Gps2, Mars & Macro. (2015). *C-booc – 20 years c-base*. New York, NY: Kickstarter.

Hegelund, A. (2005). Objectivity and subjectivity in the ethnographic method. *Qualitative Health Research*, *15*(5), 647–668. doi:10.1177/1049732304273933 PMID:15802541

Kickstarted: Finding Space and Making a Makerpace. (n.d.). [blog post]. Retrieved from https://makezine.com/2012/10/15/kickstarted-finding-space-and-making-a-makerspace

Kirk, J., & Miller, M. L. (1986). *Reliability and validity in qualitative research*. Beverly Hills, CA: Sage. doi:10.4135/9781412985659

Manzo, C. Ramella, F., (2015, December). Fab Labs in Italy: Collective Goods in the Sharing Economy. *Stato e mercato*, *105*(3), 379-418.

Menichelli, M. (2014, December 10). Come Costruire un Fablab [How to build a Fablab] [blog post]. Retrieved from http://www.makeinitaly.foundation/come-costruire-un-fablab-in-118-slides-presentazione-al-fablab-torino

NDSM. (2017). *Werf Continue as Rugged Sister to Amsterdam's Museum Plein.* Retrieved from http://www.ndsm.nl/en/

Redler, E. (2014, May). Das modelproject Haus der Eigenarbeit [The House of Self-work model]. Retrieved from http://www.hei-muenchen.de/was-ist-das-hei/das-modellprojekt-haus-der-eigenarbeit

Sapelli, G. (1994). *Economia, Tecnologia e Direzione d'impresa in Italia.* Torino, Italy: Einaudi.

Scott, J. (2011). *Hackerspaces. The Beginning.* San Francisco, CA: Internet Archive.

Chapter 2
Peculiar Elements of a Co-Manufacturing Space as Perceived by Members

ABSTRACT

The chapter focuses on the reasons that bring different kinds of independent workers to establish their activities in a shared workspace and presents their values and the personal and professional advantages they perceive in being members of a co-manufacturing spaces. The chapter presents most of the themes directly from the vivid voice of people who work in these spaces and is the longest chapter of the book. During the research, some key elements came out of the interviews and the direct observation of what used to happen in the different co-manufacturing spaces. These elements covered topics that came from the approach of people to work and life, how they used to organize themselves, how they used to act in order to gain skills, and to gather help. This chapter summarizes the most relevant aspects that came out from the research and that are at the base of the multifactory model.

INTRODUCTION

As underlined by Focardi & Salati (2015), the people interviewed showed a high level of personal commitment, a medium-high to very high education level, and they usually came from diversified previous work experiences.

DOI: 10.4018/978-1-5225-7089-9.ch002

The key elements in starting an own independent business were for all people interviewed the sense of freedom and the pursuit of self-fulfillment while creating their own businesses.

As the Authors stated (2015), these people are used to face challenges as opportunities for growth. They believe it is important to find satisfaction in what they do, regardless of the economic return, which they usually relegate to second place with respect to the return in terms of quality of life.

Their businesses cover digital enterprises, traditional companies, artisans, smart factories, freelancers, and the knowledge carried by these subjects is wide. Many of them come from very diverse and heterogeneous professional paths and this means that there is a wide range of skills and knowledge available from other people who are part of the same environment.

However, all of them are reluctant to be framed in any particular model. Officially, they are entrepreneurs, craftsmen, professionals, but when they work together they continuously mix roles, areas of expertise and methods.

Thus, the authors decided to refer to these people as "Crowdworkers", as they usually share and mix skills and resources among themselves and their products and services come from a continuous confrontation with others.

Under the right conditions, companies start to think of the resources available from others as if they were internal resources. This makes it possible to meet their customers' need for flexibility without the costs and problems of searching for and coordinating external unknown suppliers.

While mutual trust grows, further steps are to incorporate others' ideas and methods, and to share skills, but also equipment and tools, for instance letting others use these tools when not in use or during the night. For example, at Bigmagma Multifactory a new company (Makerfish) could start its activity using the hydraulic press from another existing company (Garbage Lab). Most of the exchanges, both of skills and equipment, are ruled by agreements based on mutual exchange and often don't involve monetary exchange and this is a mindset highly coherent with the basics of Sharing Economy, as companies have the availability of resources, without the possession.

However, the business model behind the multifactory model (and many other shared spaces models) is miles away from the most popular web sharing platforms, as airbnb, UBER, blablacar. Collaborative spaces are not self-driven, they need a more complex social structure, a model to follow and supervision. Even if some believe that the gigantic sharing platforms are not challenging the existing economic system, as stated by those that identify them as "Platform Capitalism" (Srnicek, 2016), it is interesting to focus on a common ground they share with co-manufacturing environments, when it relates to social aspects. The real revolution made by those online platforms are the social consequences they indirectly activate while doing their business.

But, how can a crowdworker think of the resources of others and use them as if they were its own resources? How do crowdworkers share benefits gained with different contributions from others? These are questions the authors will try to answer through the development of this book. At this point, it can generally be said that in the authors' approach what really matters is the community. These new economic relations work well when co-manufacturing spaces are run as community projects. The Multifactory model suggests a community as a managing entity of the space; this makes it possible, for example, to avoid that members self-perceived as "users", and when they start to feel the space as their own space, a new kind of community is established. Fortunato from R84 Multifactory in Mantova explains with simple words this personal sense of belonging:

At the early beginning of the project I probably did not understand the meaning of funding an Association to manage the space, and now I can say Giulio was right when he said that there is a big difference between having a space managed by someone else and having it available as a member of an Association. I was surprised because it was a feeling I never experienced, even if I've always managed public goods for work reasons, this situation is still different, when it belongs to everybody or to nobody is different, you feel it more yours. You feel a sense of belonging that is healthy. That means that you enjoy to check if all windows are closed and when I step out in the evening to check if there is something strange, not only for my space but I check also others' spaces. And I'm happy to do to it. This is definitely a positive feeling. (Andreani F., Personal communication., June 29, 2017)

COLLABORATION AND SHARING

Collaboration is a funding element of this new way crowdworkers intend economy.

There is no doubt that companies always used to collaborate, and one-to-one collaborations have always been very common. A traditional form of collaboration is the creation of a co-financed and co-managed entity to provide category representation and shared services or to combine efforts to buy together very expensive means of production, as it often happens within agricultural enterprises.

But collaboration that takes place in co-manufacturing spaces is different, as companies which are part of these workspaces use collaboration as a way to create a shared repository of assets (means of production, skills, knowledge) which can be used by every other company that's part of the system.

Every company makes some selected resources available to others, and the whole of these resources become a sort of collective value of the community of companies.

Traditional collaboration between companies did not involve the transfer of knowledge, everyone was willing to keep its secrets safe about what to do or how to do something. What was shared was the client or the machine.

People like to interact with each other, they can work in an office but as soon as they need to take a break they the go out and they start talking with another member about what they are doing and suddenly they came up with new ideas. So why don't we share our knowledge our resources and we create something better. This new way of cooperation instead of competition is, in my opinion, the future. So that's how these kind of coworking spaces are changing, a new working style. (Toriz Morales, M., Personal Communication, August 24, 2016)

Crowdworkers are proud to share what they know with others inside their community and are more curious about learning new things from others:

They come for using tools of for the events but then they meet each other. They come here and they do their stuff, but then this is a way to talk to others. They are not very special people but are open minded. Open minded people, open to new stuff and to new experiences. Working together is a new experience, not in a normal way, not with concurrence but in collaboration. (Roeder, M., Personal Communication, June 10, 2017).

As stated by Focardi & Salati, (2016) a crucial point for new and small Companies is to be fast: "There used to be a time when it was very important to protect an idea, let it grow internally in the company and then to develop an exclusive product/service that could last for years.

Nowadays, the life cycle of ideas, products, and services is so fast that it's less important to protect them. It's much more important to develop them as fast as possible, in a very reliable and flexible way, to target the market as quickly as possible.

A single small or medium company, an artisan or a pro should have many good ideas throughout the years, and they can be very different from others, so they can't rely only on their knowledge, skills and experience, but they need an easy, fast, reliable ways to get competencies.

If they're part of a system where they're used to an everyday free exchange of knowledge between well-known subjects, they can find the resources they need, saving financial expenses and avoiding the cost of retrieving resources (for instance, to obtain a loan), and the cost of creating a completely new coordinating structure for each single project."

Most of the Companies that are based in Co-manufacturing Spaces in most cases are startups and are small economic agents like solopreneurs, SMEs, artisans. They have most of the technical skills needed to start and run their own business and to develop new projects, but they miss many other skills and lack many resources. For example a craftsman can have deep skills on how to work wood, metal or leather but can be totally lacking on social media or a young designer can have many skills on graphic design but can lack on business management like how to deal with local regulations and laws.

Shared Workspaces allow them to retrieve these resources from other members of the same environment.

MIX OF SKILLS

From heterogeneity comes a multitude of skills, points of view and experiences. After a first stage of orientation in an unusually heterogeneous work environment, this is the opportunity to find inspiration from diversity: "Here we all do different things, but we collaborate, we exchange opinions, it's a collective effort" (Oliveira, F., Personal Communication, May 3, 2013).

I had a shared studio before, but we all worked in the same field. It didn't work because I felt I was stuck in a pattern we were all people sewing, but I needed more ideas, to be more connected with people coming from different worlds. (Evasé, M., Personal Communication, May 3, 2013)

Practical knowledge and theoretical knowledge are placed exactly on the same level and the comparison between the points of view results in a mutual increase:

While talking to the architect firm in front of my workshop I discovered many new things about products design, as the objects size, that an artisan might only guess. Of course, I had my own ideas and followed my own empirical rules, but thanks to these architects I discovered that object sizes had already been standardized. The same happens from my side, it is an exchange between my years of practice and their theoretical approach. As I have learned something from them, I have also given something in return: when you design something without having any clue on how to physically build it, there is a risk that then the object brakes. This biunivocal exchange of skills is very nice, the fact of working in synergy surely helps both parties. (Merli, O., Personal Communication, November 2, 2012)

This is only possible if there is an explicit interest in finding heterogeneous professional profiles, which has to be a point of interest when searching for new members: "We like to have different people inside the space, it's better to learn from different people and to share different points of view, but it's a very free process, we meet new people and then they join us." (Carvalho, F., Personal Communication, May 5, 2013)

Very often, the interest in the activities of those who stand side by side is beyond the simple need to satisfy their own specific need, but is of a more general nature and this allows for a professional cross growth driven by opinion exchanges, advices, questions: "There are a lot of different skills available here, so I've learned so many things that have nothing to do with my job but that can be somehow useful" (Merli, O., Personal Communication, November 2, 2012).

The heterogeneity of skills and views is one of the main advantages of working in a shared workplace, as the opinions of those who do a different job broaden views and introduce innovation elements from which unexpected combinations of products and services can arise:

"I think I'm really lucky that I can work in a space were everyone is different and has got his own specialty and I can learn from them" (Lapo., Personal Communication, May 3, 2013).

This is a shared studio, I was in a sewing studio, but I didn't like the way it was organized. I've been here for six months and this was a big improvement. I did more in these six months than in the last one year and a half, it's a different dynamic for me. It's good in a funny way. (Evasé, M., Personal Communication, May 3, 2013)

MIX OF NEW TECHNOLOGIES AND TRADITIONAL KNOWLEDGE

This diversity also includes the coexistence and integration of skills, methods and models of very distant professional worlds.

People in practical jobs can find hints to grow from other members. The traditional figure of the young apprentice is replaced by this mutual help between professionals and artisans, the creative reuse of previous skills and the help of technology as a mean to learn new skills.

Real technology is one that helps the human being to simplify everyday life and must be for everyone. True technology must be for everyone, not for an elite. Then we use sewing machines that have been the same since hundreds years, but we use a lot of communication technology, both for customers and for finding suppliers. The web, for those who start a new activity without any specific experience on that field, like in our case, it's an incredibly huge benefit. (Morosi, M., Personal Communication, October 10, 2012)

Some companies focus their core business on technological innovation: "It is what allows us to work, because without the production technology and sales technology we could not do what we do" (Bocola, M., Personal Communication, October 1, 2012).

While others consider technology something completely inaccessible and rely on the proximity network to bridge their technological gap:

- "I have many difficulties with technology, I get lost also if I have to set up my phone" Cristiani, E., Personal Communication, October 17, 2012).
- "We don't match so much with technology. I'm not so much a technological person, mine is more a manual job. It's more a matter of craftsmanship" (Woolflex, L., Personal Communication, January 22, 2013).

In the current professional world, it is important to know how to respond speedily to customer requests, which can change very quickly. The characteristic times of renewal of skills are speeding up and personal obsolescence is increasingly lurking. Getting in touch with a diversity of professions offers constant stimuli for professional upgrade and reduces this risk of being outdated:

- "I do not follow any courses, in my field more than having professional upgrades it is about pursuing a personal enrichment" (Saiu, F., Personal Communication, October 20, 2012).
- "Continuous self training, for me now it's experimenting with new materials, new forms, new techniques." (Cristiani, E., Personal Communication, October 17, 2012).

COMMERCIAL AND MARKETING ASPECTS

Managing business and marketing is undoubtedly the most delicate and complicated part.

All people encountered during the research agree on the need to communicate effectively the meaning of their work abroad:

"Communication is an important thing in which I believe. And my pieces should speak for me" (Saiu, F., Personal Communication, October 20, 2012).

- "Communication is a way to get to others by proposing something right and focusing on what you communicate" (Cristiani, E., Personal Communication, October 17, 2012).

In general, crowdworkers are able to think, design and create functional, useful and innovative objects or services, and many recognize the importance of feedback from outside to grow their own project:

"Every piece has to communicate something, and I have to communicate something about that piece too. I'm slowly discovering myself through my work" Saiu, F., Personal Communication, October 20, 2012).

For almost everybody, the real difficulty is when the time comes to get on the market.

"We are able to create new things. We have many ideas, but then the hard part is to bring them to market" (Carvalho. F., Personal Communication, May 5, 2013).

Selling is a critical point for any company, but innovative companies also have the added difficulty to communicate to the public the feeling and value of a new model and architecture to develop products and services. This takes additional consumption of energy and time.

When we started our Tool Library it took a little while to explain what we are gonna do. We wanna let tools out to people and most of the time the response we got from older generation was "your tools gonna get stolen" or "what's gonna happen to all your stuff when it brakes." For us things do brake. We try to maintain them, to fix things. We try to be planned, up to licence, which is like a bit capitalistic. (J. Shar, Personal Communication, July 23, 2016)

They recognize the lack on the commercial side, but they are also aware times are difficult for making business:

- "We need to develop the business part. We need to sell and know sellers and business, but that takes time. We are creatives, and we do not know how to sell" (Carvalho. F., Personal Communication, May 5, 2013).
- "I understand that with the crisis that is now there is a need for a good melange of patience" (Saiu, F., Personal Communication, October 20, 2012).

Focusing on the quality of products and services offered is a common choice among crowdworkers but more for personal satisfaction than for strictly commercial reasons.

Satisfaction is the people who come to tell you that your product is an excellence, you have to do the production as if it were for yourself, but you do it also for others. I do not put chemicals into something I should eat. It is a business principle: I produce something as if it were for me, but I also sell it to others. (Ricucci, A., Personal Communication, January 26, 2013)

On the other hand, this peculiarity, the fact of rethinking the production system from the base, can become, if well exploited, a selling point.

Again, being part of a unitary system, the shared workspace, can be a way to find support and build common growth opportunities.

A concrete example that has appeared several times in the Bigmagma multifactory may be attendance at fairs. As different companies take part in the same fair, it is possible for them to make a series of joint actions structuring on multiple levels.

The first level is the opportunity to present to the event organizers as a single entity, specifically Bigmagma, getting discounts and/or facilities.

The second level of sustainability is to be able to think about assembling products in a joint way and internally to the workspace, reducing costs, times of accomplishment and obtaining a custom product.

The third level of sustainability is to plan trips together, minimizing costs and maximizing time / work performance.

The fourth level of sustainability is the presence of other trusted people, which allows to preside over the spaces.

All this makes it possible for a very small company to access prestigious visitor opportunities, overcoming the costs, logistics, and organizational difficulties.

BENEFITS OF SHARING KNOWLEDGE

Among crowdworkers, the culture of sharing, the open knowledge and the open source is something that is well identified and visible. To foster this culture of sharing knowledge is not about sharing technological equipment, but the habit of sharing in itself.

- "This is an open space and open space means also open knowledge. People should try to transfer their knowledge to other people" (Parade, M., Personal Communication, July 15, 2013).
- "Collaboration is to integrate art into a household object or people who have the ability to use other materials that work with the jobs I do, so it is something I look for, welcome, helps a lot, working together is beautiful" (Saiu, F., Personal Communication, October 20, 2012).

The best cases are where the formal or informal coordination of the space can support and promote the cultural activity carried out by the members, both outward and in-house, and is able to involve also the most traditional companies. In those cases potentialities of the transfer of culture can express themselves and allow all the actors to benefit from it.

We like to promote peer-to-peer learning, we work with universities abroad and they send us their students, they come to study inside our facility instead of going to university, they actually look for a non-academic environment and instead of teachers we pair them up with members. It's not exactly coworking, it's not exactly a school, it's a kind of a new environment for learning. (Tham, C., Personal Communication, September 18, 2013)

As stated by Focardi and Salati (2015) "At first sight, the return should be intended as a mere exchange of professional services, a way to offer and get services without money, or a kind of barter: "I have a need, I looked for a skill, I found it, I offer something in change."

This is something that happens, but there is something hidden much more interesting, that's the capa- bility of the system to increase through knowledge sharing the opportunities of growth of the companies that are part of the system.

What happens is that availability of competencies enables and leads people to think in a different way. They know that when they face some problems, or professional challenges, they can count on a wide range of capabilities and skills that are available and reliable, and this constant, continuous availability of resources enables entrepreneurs to think in a wider way.

It's a buffet of skills. Entrepreneurs know which resources are available and how to get them. Maybe they don't need them for a long time, but they are aware of the availability and this leads them to take on big projects because they know that they will have access to all what they will need.

So, when they put knowledge into the system, what they're doing is raising the global potential of the system by itself, and if everyone in the system offers something different, the result is that every node in the system can grow faster and steadier.

It's a way to build a system where all the players give their contribution to increase the capabilities of every single node of the system itself."

In Co-manufacturing Spaces most of the investments are on the human capital, as positive and collaborative dynamics are the real assets of companies based in a shared workspace and this requires people to be willing to share and collaborate.

We help each other, it's a culture. We help each other, for us it's normal. It is based on honesty. Every person is not only a worker but a human being. We cannot help everybody, but we can try it and if we can't, maybe we know people that can help. Not all is possible here but maybe in our network there are places or people. (Roeder, M., Personal Communication, June 10, 2017)

Of course, this makes it important to carefully find and select the members, as these dynamics are activated if all people working inside the space are open to change, to new trends and appreciate the value of sharing.

We welcome also people with business ideas but this is not place for classical ideas, for fancy new and hipster startups, they work with an old business model and it is always the same and it doesn't feet with the Konglomerat. (Roeder, M., Personal Communication, June 10, 2017)

Since the moment people join a workspace, from the moment they sign the contract they are not renting a space, but become members of a community of workers. The primary goal is the mutual help between entrepreneurs, the way is to share personal knowledge with others inside the association. The sharing of a physical space becomes an ancillary element and not the primary one.

Nevertheless, the idea of sharing is not already widely spread among independent workers. Many entrepreneurs are not used to share their know-how, craftsmen are traditionally very jealous of their knowledge, and sometimes they don't even teach their apprentices their professional secrets. Most Co-manufacturing Spaces are aware of this and design their internal processes in order to foster the spread among their members of ideas and practices based on collaboration and sharing, through governance solutions based on collective management of common goods. This is the case of the association *Konglomerat* in Dresden.

Here people come mostly for their free time, we experiment new ways with small projects as for example now we got a donation of 5000 euro now it's a question of how to share that money. And it is an experiment and experience for all of us to do it fairly. So they experiment with a small project. It is not only for the administration of the space but we try to think what good could be done with it. We use a software that is called cobudget, it is like an internal crowdfunding. We collect ideas and then we discuss inside the community. We could use that money for example for buying some machines, organize a festival, give to people that don't have the money for the membership, these are the ideas emerged in the plenum. (Roeder, M., Personal Communication, June 10, 2017)

This is a way to develop some tools to let crowdworkers be more confident in sharing their knowledge also in a business environment.

Of course, a new system requires new tools and there may be many ways to develop them, but the key point is that among Co-manufacturing Spaces there is a changing idea of what the primary goal of running, hosting or promoting economic activities is:

Traditionally, entrepreneurs go for profit and growth. The one thing that is important is profit, it's all about growth and profit. But what we are trying to set here is not the company's growth, but the growth of people within it. (Nichols, A., Personal Communication, February 27, 2013)

BENEFITS OF SHARED WORKPLACES

A Way to Let Ideas Grow and Spread

This cultural imprinting for sharing is also reflected in ways to innovate and develop new products, services, projects or solutions. It is possible to activate many companies in parallel, at the same time, on a single project. The most obvious advantages are: a widespread investment and an already experienced co-ordination.

There are not so many places in Germany like FreiLand when you have in a very small place, a relatively small place, so many people with a creative brain and you have short way to go around the corner to meet a musician, or a guy who has a music studio to ask them some suggestion for building a new speaker phones. Or you can make a short walk around and talk to some people and then you make a new project. (Parade, M., Personal Communication, July 15, 2013)

From contamination among nearby companies, arise experiments and collective re-elaborations of different experiences:

An example of collaboration are the Bigmagma lamps, for example, which merge different techniques: hand-cut glass, CNC-cut for the wooden base and laser cut for the plexiglass part. At the end: different materials, many of them are recycled, cut with different techniques, by different professionals in the same space. (Ricca, E., Personal Communication, October 1, 2012)

In some cases, these collaborations take shape as examples of circular economics, such as a simple and immediate reuse of production waste from a nearby company:

Our zips that we would scrape out. We gave them to Artedi' jewelry atelier that turned them into earrings. And now we sell them in our mono brand shop. And also the outfits that Lascialascia (an architect firm) builds up for our shops. (Morosi, M., Personal Communication, October 10, 2012)

In other cases, the stimulus from others leads to a creative and unexpected reuse of pre-existing skills:

"One day you make a thing, another day you make something else, but I can use the same stuff I learnt at University to have a different perspective on what I'm doing" (Oliveira, F., Personal Communication, Mar 3, 2013).

For Evra I realize very particular and quite complex filters, which require a specific fabric. We hadn't done anything of the sort, but we had a look at the sample, looked for the right fabric, tested and studied the right processing. At first we did not even know if it was possible to do it. We have learned that we must never say "we can only get here". You can always go a bit further. With this experience we understood that you can always move a bit to other fields. (Woolflex, L., Personal Communication, January 22, 2013

Or the positive competition has as consequence of observation and influence of neighbour's actions:

There are people doing different things here at Oficina Colectiva, and we're very different. They have stuff we don't have, because they're more rational, more organized than we are here, upstairs. Downstairs they're more logical, and they think much better than us. And this infects us in a good way. I think we both inspire each other in some way. It's really good to see the reactions: if we start putting post-its on the walls, they first react a little bit strangely to that, and after a month you see they're doing the same thing. If they start doing commercial work, we think "damn, we should do commercial work also" and we start following them and they start following us. it's a kind of a really healthy competition when they do something we have to positively react and do something good, and it's the same for them. It's a kind of creating that feeling like "let's move, let's move let's move, let's do stuff, let's do things" This is the kind of feeling I feel here. (Quinta, R., Personal Communication, May 5, 2013)

A key point is that in a shared space the circulation of ideas is part of the daily life and happens naturally:

Many collaborations take place heres, BIGMAGMA has made for us a marvelous and indispensable eye-catching installation that has led us to exponentially enhance our business appeal. Then other collaborations, banally joint participation to fairs, sharing costs and organizing outfitting and a common logistics. Then in the everyday activities there are a thousand examples, perhaps banality, like exchanging opinions about suppliers, or the development of a small solution that is based on an intuition of a neighbour and which alone would require days of research. Or a reciprocal exchange of tools and machines. You don't even perceive them because they're so rooted into the daily routine. (Bravi, D., Personal Communication, October 13, 2012)

But to be able to transfer their skills to others and to be able to receive them is not something that can only be accidental, it is necessary to activate specific dynamics:

"Right now, no, I do not collaborate with anybody, but it is something I want to do. At Mage there are very good craftsmen, there's a very good carpenter, who produces furniture and we are thinking about how to add ceramic inserts" *(Cristiani, E., Personal Communication, October 17, 2012).*

"Between similar businesses, you could collaborate by creating a common commercial effort by contacting stores and then making them a joint proposal" (Cristiani, E., Personal Communication, October 17, 2012).

This, however, even if it allows a confrontation on everyday issues, often is not enough to lead to structured collaborations: there are opportunities to facilitate the exchange of views and the circulation of knowledge. These occasions are often created by the companies themselves, while sometimes defined by the initiators, which have a long-term vision.

"We clean the space by ourselves, we bring the food here together, we are residents, everybody has the key but we don't have special rules. At Friday lunch we meet all together and share ideas and opinions. It works" (Carvalho, F., Personal Communication, May 5, 2013).

This deep mutual understanding between members of the same workplace helps entrepreneurs to solve the problem of managing innovation, which Chesbrough (2003) clearly underlines when he states that "SME need to reveal and promote many new ideas, many technologies and many plans, even before they can start actions to protect them". Small economic actors need to reveal their secrets to foster resources, but while revealing them they vanish part of their competitive advantage.

In Co-manufacturing Spaces people are free to talk about their plans and assets, as there is mutual trust. Many people have many doubts that this openness and mutual trust are possible in business environments, the general idea is that competition always rules.

It is clear that this kind of collaboration is something completely new and requires a total change of mindset; it is not easy for anybody, both traditional and new companies, to understand how this system can work. In this new environment ideas are locally shared, but are protected too in a new way, using new tools that replace the traditional patents. This change of economic paradigm requires, it is clear, new tools too, some specific practical tools that can be used by crowdworkers to make it working. Some examples of these new tools will be given in detail later in this book.

The type of access to resources made available by other companies within Co-manufacturing Spaces is somehow similar to a buffet of skills. Entrepreneurs know which resources are available, how to get them and from whom. Maybe they don't need them for a long time, but they are aware that when they face new problems and they don't know how to solve them, there are others who can directly solve them, or help them to autonomously find a solution.

This availability of resources leads entrepreneurs to start or join challenging projects, as they know there are reliable and accessible resources of skills and help around them, and this constant, continuous availability of resources enables entrepreneurs to think in a wider way.

So, entrepreneurs put their skills on this shared skills repository, and take what they need from it. While planning and running new projects they use and increase these skills, which will be put back in the repository of skills.

In this way, the skill level of the whole system grows and increases the overall capability and the global potential of the system by itself. The result is that every node in the system can grow faster and steadier.

In seeking to make their businesses sustainable, companies that are part of shared space test new ideas, production modes, and processes. The daily attempt to propose exclusive services or products leads to experimentation and hybridization and the consequence is product innovation. In this sense, the concept of "innovation" is closely linked to the one of "sustainability":

"When you find a new way to see something and to operate it, this is innovation for me" (Parade, M., Personal Communication, July 15, 2013).

In general, the research of sustainability is not so much seen as a set of practices, but as a background that inspires the overall business activity. Antonio, a farmer form the *Microfiliera Pollino* (ITA) says:

"You have to be first organic inside your head, then you can put it into practice" (Ricucci, A., Personal Communication, January 26, 2013).

So Co-manufacturing Spaces can foster both economic sustainability and innovation at the same time, swapping the concept of capital from economic capital to be used to buy new skills and equipments, to a capital of relationships that gives availability of equipments and brings skills to use when needed.

John from the *Station North Tool Library* in Baltimore (US) is conscious of this role:

I think we are showing people that there is other ways to do things. There is a healthy way to do things and to build communities and to collaborate with other organizations, spaces and people. We are living in a planet that is destroying itself by just producing more, and more, and more things. So what do we do? We show that you don't really need to own everything. A large community could own the essential tools that they need, take care of things together, work together, things like that. (J. Shar, Personal Communication, July 23, 2016)

Commercial Strategies

The most innovative part is how this exchange, and also the mere availability of resources, can lead to a growth of the system as a whole and of each element of the system.

This growth follows two paths, one that is exclusively internal to the closed system of the companies working together, one that is open to the outside.

The internal path is a sort of circle that takes place between companies. The availability of skills brings them to explore new business opportunities, which lead to new projects, where skills from different companies are mixed together. This way, all the actors involved in the project put some of their skills, giving value to the project, and at the same time learn from others.

Usually, in Co-manufacturing Spaces, many companies choose to use other companies that are part of the same space as suppliers, whenever possible. This is both for convenience and for minimizing initial capital investment, as it often happens that non-monetary remuneration mechanisms are triggered, as well as exchanges of services, machinery and skills or barters.

Social Capital is a system of relationships. When certain skills are needed, but these cannot be paid for by cash, one goes back to the barter tradition. This formula is very interpersonal and cannot be dropped from above, can only be facilitated. Therefore, even when it is said that the Mage is a space with rules that must be maintained between the property, the usufruct and the subject manager, these rules are wide mesh. (Inti, I., Personal Communication, October 23, 2013).

As far as the identification of their reference market is concerned, the services and products offered by the various companies are very different, so are the customers.

"Our customers range from small, to very small. In fact, there are also companies, design studios, but I like to give to anybody the chance to lasercut their project" (Ricca, E., Personal Communication, October 1, 2012).

Many companies inside shared work spaces are small and in many cases are even run by only one person and it is quite common they incorporate many of the craftsman's typical operating models. So, in general, the products and services offered are mainly aimed at the local market. Despite this, many of them identify their business as global:

"My business is both local and global. It's local because I work here in Lisbon, but it's global because I have working connections with people from all over the world" (Evasé, M., Personal Communication, May 3, 2013).

Being part of collective online sales tools, such as a common facebook page, and other social networks, allows companies to receive orders from anywhere in the world, and this, although usually a marginal part of revenue, is still a starting point to grow from.

I'm starting now to sell my works around the world. It's nice to see that my work starts to go around the world and in other countries. I have got a blog that I started as a joke because I started working on my illustration on my lunch time that's why I called it "No time for lunch". People find me on my blog, maybe it's word of mouth. (Lapo., Personal Communication, May 3, 2013).

It is recurrent that crowdworkers are travelers too, many of them are or started as freelancers and have always been used to move a lot because they could work from anywhere. *Poligon – Creativni Center based in* Lubiana inside a former tobacco factory is used to host nomadic workers:

They travel. Because it is very nice to travel. Everybody loves it. If you only need a computer and a wifi, why stay always in the same place? Why always stay in the same space. You can meet many travelers, they just say "in the last three years, I've been traveling." If they feel really comfortable in Poligon, they will come back in one or two years. (Toriz Morales, M., Personal Communication, August 24, 2016).

This openness to other worlds helps to offer their activity to a global marketplace:

"It's a global business. We have here people from all over the world, the production is local, but we sell it globally" (Carvalho, F., Personal Communication, May 5, 2013).

I'm an architect. I have an architecture studio I created with two partners. I have many collaborations in various parts of the world. The company's headquarters are in Barcelona. We are free enough to move and make other collaborations and this is a way to enrich ourselves professionally and the fact that we are in a space where there are people who have a different profession is a further enrichment for our work, and this is also the beauty to in a place like this one. We did not do real work with the illustrators except the exhibitions and the Balneario's Open Day but their presence influences us indirectly. (Brazao, R., Personal Communication, May 5, 2013)

"We sell only online, which is both the most universal and the more organized way to manage the orders we receive everyday" (Ricca, E., Personal Communication, October 1, 2012).

One of the marketing strategies adopted is direct sales by manufacturers to consumers. This eliminates all fixed costs relative to intermediaries and internal administrative figures:

We tried to cut brokers, agencies, representatives. We want to get our customers directly. We want to produce the products and sell them directly to the end customer, so we are opening mono brand stores. Then we have some trusted resellers and the foreign market, which is one of our basics incomes because it pays on time and it is sure it pays you. (Morosi, M., Personal Communication, October 10, 2012).

Often, mutual help dynamics are self-generated when using external suppliers too. Companies not only tend to exchange information about their suppliers, those that are reliable and that fit best the need expressed, but also give direct support in dealing with suppliers. There is also an indirect effect while interfacing with suppliers as members of a Co-manufacturing Space. As members of a Co-manufacturing space, which is huge and well known, companies can easily get discounts or special offers. From the supplier perspective, single customers are small but the Co-manufacturing Space, as a whole, generates a volume of relevant business that deserves to be preserved.

Initial Motivation to Join a Shared Workspace: Cost Reduction and initial Investment

A shared space allows to significantly lower the initial investment to start an independent business. First, management costs are lowered because fixed costs are split between more entities, achieving economies of scale. Then, the ability to easily modify the amount of private space occupied inside a Co-manufacturing Space allows the crowdworkers to start their business with a small space reducing the variable costs associated with the physical space with the minimum needed to start.

"She has won a quiz on a TV show and instead of buying a home or a car we have decided to invest in our future and to open a company" (Bocola, M., Personal Communication, October 1, 2012).

"The funds to get started are coming from the fruit of my work, I worked in many different situations and I'm reinvesting on myself, with the skills I have" (Saiu, F., Personal Communication, October 20, 2012).

"My business does not require a large initial investments because I use few equipments, so I dipped into my savings" (Cristiani, E., Personal Communication, October 17, 2012).

In addition to this, within a general cost reduction strategy, crowdworkers often resort to friends and acquaintances. This helps to reduce start-up costs.

Although they often have to take on various kinds of risks, crowdworkers do not perceive this condition of economic difficulty as a limit, but as an integral part of the challenge undertaken.

In 2007, for our first trade fair, we did not have the money to pay our booth. We paid it three days later, going to the bank and pouring the cash collected in the first three days of the fair to cover the check we had given to the event organizer. You can risk, when you believe in something. (Morosi, M., Personal Communication, October 10, 2012).

"We started with little money and some basic machinery" (Bravi, D., Personal Communication, (2012, October 13).

It is much more educational to do as much as possible with your own funds, large or small it doesn't matter, for example I remember that there was a tender that provided 70,000 euros at a very good rate, but I'm not very much in favour of such a big loan. Sooner or later you have to give them back. A large amount of funding would make the start-up easier and maybe we could have managed it, but it's also reasonable to say that if you have 70,000 euros in your hands you spend all of them but you have no experience. Now maybe we would be able to handle such a big loan, but only after a lot of experience, work. When you know for example which fair fits your product or which market does not work. (Colombo, S., Personal Communication, October 13, 2012).

Moreover, a Co-manufacturing Space relies on communication tools, such as a website, contacts with the local press, a facebook page and other kinds of social networks. Therefore, a new company can start its own marketing business using at first the existing communication tools belonging to the Co-manufacturing Space. As a relevant consequence, it can have much faster access to the market.

Finally, for those who start an activity from scratch, the presence of other existing companies is an advantage because it allows new crowdworkers to ask specific questions to others before entering the market. This allows *a priori* verification of the soundness of their choice and gives a huge helping hand having feedbacks from others during the following phases of development of the activity.

Clearly this approach to work doesn't fit everybody and can't be understood by everybody. To start a new Co-manufacturing space pioneers are needed that can foresee these opportunities but, in any case, these spaces are addressing a completely new kind of entrepreneur.

Most of the advantages of being part of such a space are not immediate but are discovered by the crowdworkers after a while, and they experience them by carrying on with their daily working activity. Most of those interviewed couldn't guess most of the advantages in advance, before having a first-hand experience. This is also because most of the reasons why these constitutive elements of a co-manufacturing space become advantages are counter intuitive.

"Everything started by chance, we were looking for a place to develop our profession, then we found this great space, we got together, other people came after, and then our small community started" (Brazao, R., Personal communication, May 5, 2013).

Often, the initial motivation to work in a shared work space is far more pragmatic and simple, rather than focused on complex processes:

I didn't want to work at home or isolated because I know I wouldn't have been very productive. Any working environment I think influences a lot your creativity, your health, and your well being. People are an inspiration and the space is very nice and this has a big effect on your creativity. (Lapo., Personal Communication, May 3, 2013).

I had no expectation, I hoped that our cooperative could find a place where to stay: Maybe to have a support to grow especially in those aspects where we felt more shortcomings such as in communication. But in general, I wanted to find an availability. All things that I see that here slowly are happening. (Scattolini, C., Personal Communication, March 5, 2018).

Others were just looking for a solution to the uncertainty of the labour market:

"There are things to do, but what's missing in this village are people. Someone is coming back home, because of the crisis, and when you lose your work, at home is simpler" (Lista, A, Personal Communication, January 15, 2013).

Some of them decided to establish their activity in a Shared Workspace simply because they wanted to find people similar to them and not to feel alone:

"I've read articles on the web, about a creative space full of shared energy and I wanted to be part of it" (Cristiani, E., Personal Communication, October 17, 2012).

"We are at Mage because we wanted to find people with our style, with our patterns of reasoning, we wanted to confront, we wanted to draw from the experience" (Morosi, M., Personal Communication, October 10, 2012).

"I had been working alone for one year, but I was too bored, I hate working alone, I need a place to share ideas and to talk with other creative people" (Duarte, N., Personal Communication, May 5, 2013).

The expectations that the crowdworker had before starting to work in a co-manufacturing Space in general terms were fulfilled:

"Mage could meet my expectations, which were to find an environment open to exchanges" (Ricca, E., Personal Communication, October 1, 2012).

My expectations were all met, my idea was to come and work and find activities with which to exchange ideas, emotions and stimuli. For me, it is an opportunity to do my business with low costs, but also to exchange. Even just a simple chat, an advice. If I were in a beautiful shop, in my own workshop, but alone, I could not have these cultural exchanges. (Mancini, S., Personal Communication, November 7, 2012)

In some cases these are more strictly related to a personal challenge:

"My expectations are on myself, and I still have to give something to this town, in Milan" (Saiu, F., Personal Communication, October 20, 2012).

In other cases, they are only partly fulfilled. One of the reasons is because the co-manufacturing space was initially born with different assumptions:

Mage at the beginning was presented also as a kind of business incubator, paradoxically one could expect that there would be someone who in the long run would want to contribute to our company, a business angel. At least we were expecting some expert who provided advice and experience. (Colombo, S., Personal Communication, October 12, 2012)

"We thought we could find people working in the same field, but this hasn't been the case. Anyway, something nice and different was born" (Morosi, M., Personal Communication, October 10, 2012).

In other cases, there was no expectation related to the sharing because some of the case studies were born in the absence of a long-term project.

We decide to work by ourselves first of all because there are no jobs, it's very difficult to find a good job. So we decided to start our own business, which makes it possible to work on architecture and to do things. We didn't think if it would have been possible, we just did it. (Carvalho, F., Personal Communication, May 5, 2013)

"I wanted to have an income to maintain my family and at the same time to carry on with this great passion I have" (Cristiani, E., Personal Communication, October 17, 2012).

And therefore the choice to share a space with other businesses was simply the most immediate response to other needs:

"This was not something we had in mind at the beginning, something we were looking for, but we were really lucky to find it" (Quinta, R., Personal Communication, May 5, 2013).

Often, however, it is only after some time that is possible to realize the real benefits of sharing with other entrepreneurs. These new needs become even more relevant than the initial ones:

At the beginning I was working at home and when we started to have to show our work and to receive clients we desperately had to move out from home because you can't receive clients at home. So we found this space really easy and quickly. What really attracted us at first was the space, but after a month or so we started to feel we wanted a place that was more than just the space but also the people. And after a while I felt was like a second home. (Lapo., Personal Communication, May 3, 2013)

"Positive things are far more than the bad ones, [and are] linked more to the aspects of coexistence of so many people all together in the same space. For me, however,[it] was positive because I learned to live with people" (Merli, O., Personal Communication, November 2, 2012).

At the beginning I was a bit afraid, but now I think it was the best decision I could take. I'm the only one sewing here. I get ideas from seeing other people working, also if they are not doing the same thing. This brings me ideas. For instance we've people making bicycles here, so I had the idea to make bicycle bags with recycled materials. (Evasé, M., Personal Communication, May 3, 2013)

Initially what the crowdworkers perceive as an immediate advantage is the ability to work in a larger and aesthetically pleasing space:

In Milan all the beautiful spaces devoted to craftsmen have been transformed into lofts and what's left are only boxes, cellars or dark spaces. I was looking for a nice space and to share the rent costs with someone and so I found it here. (Merli, O., Personal Communication, November 2, 2012)

Often the possibility to customize the space is perceived as an added value:

We came to see the space, and we were amazed. We could do everything we wanted, because it's a little bit rotten. Actually it's not perfect, but it's perfect for us because we can prototype, we can cut, we can paint on the walls, we paint the ceiling, be on the floor plan like kids, that's how we like to work here at Toyno. (Quinta, R., Personal Communication, May 5, 2013)

A great advantage is the good impression a crowdworker makes with his customers, suppliers or collaborators when they enter the space:

"People come here and see this is not just a sewing studio. They find architects, people working on drawings, people building bicycles, an art exhibition" (Evasé, M., Personal Communication, May 3, 2013).

There are also personal aspects, often linked to the need for reconciliation between family and work:

Yes (I'm happy) a lot more than before, I was doing a job for ten hours a day, it was very stressful, I never had time for my family and my little daughter, I did not see her for days. Here I have the opportunity to manage my time and this means raising the quality of my life. (Cristiani, E., Personal Communication, October 17, 2012)

Other aspects come as a consequence and depend largely on the specific type of community which will grow over time:

For sure, staying in a beautiful space boosts your individual activities, inspires you, you work better and you produce more, then there are some people like Lapo who did an exhibition here and from that she received other invitations to make exhibitions and artistic residences. (Brazao, R., Personal communication, May 5, 2013)

Sense of humor, I think that's really what brings all of us together. My work has a lot of it and everybody here has a little bit of it; so we feel quite open to say anything that goes through our minds and we never take it as a bad thing. (Lapo., Personal Communication, May 3, 2013)

Mage is a place that banally in today's society people perceive as a place where to go on the free time, an after-work space. While absolving also to this function, Mage brings leisure into a working place. (Colombo, S., Personal Communication, October 12, 20112)

For those activities that require almost exclusively a laptop, the presence of a real physical space allows to mentally and not only physically root the new business:

We're here because I personally worked a s a freelancer, I was used to work from home, then I came to Berlin, I started working with teams again, and when I came back to Lisbon I decided I had had enough to work from home, I was really feeling alone, so I decided I wanted to work in team, I wanted to work with people, so I started looking for a place to work. We needed a place to work, a place to establish Toyno, a physical space. It's a kind of a contract, when you have a space you have to go there, it's not only the fact you have to pay for it, but the fact it's yours, you're occupying it physically. (Quinta, R., Personal Communication, May 5, 2013)

The ability to immediately create prototypes and to see the result in real time becomes one of the benefits that gives the more satisfaction:

When people ask me where I work, I don't say "studio", I don't say "agency", I don't say "atelier", I say "Oficina", because it's part of the name, but also because oficina is a place where you put your hands on stuff and you do stuff with your hands. (Quinta, R., Personal Communication, May 5, 2013)

One of the most unexpected benefits is the activation of lateral thinking when other creatives, who are not experts in that specific industry, give their personal opinion about someone else's activity:

"Sometimes here you get opinions on your work, coming from people who don't have your specific skills, but it's not a technical opinion, they are more free to give me their opinion" (Evasé, M., Personal Communication, May 3, 2013)

Having people around is a great help, a one-to-one support because you are not isolated from the world, you are not in your labs, you see others that are similar to you and you can measure, find connections, but also simple building blocks of companies that look at it is a incentive to continue, grow, grow. When you're back in a job, you think you can never do it, see your neighbour who is in a similar situation and you say "ok, I'm not alone, then you understand that maybe you're not all wrong and you have a stimulus to go ahead in complex situations. (Colombo, S., Personal Communication, October 12, 2012)

How to Grow: Reuse of Competences and Self Education

Although in craftsmanship there is the passion for handmade products, crowdworkers often do not come from a long vertical growth path, they have not "learned the craft" by being the apprentice to someone experienced and have no sedimented specific skills.

They cover this lack of experience in part by re-using previously acquired skills in other fields:

Having worked for 25 years as a computer technician has allowed me to have a method that I reuse in my carpentry work. Thinking in logical steps allows me to apply an ordered method to a job I have never done, I could do an algorithm of my work. (Mancini, S., Personal Communication, November 7, 2012)

Or through the support that can come from the network of proximity. The proximity network indirectly feeds the creativity of individuals because everyone involved tends to demonstrate to a peer "how to do" something related to his profession and for this reason he proposes something original and high-level and recognizable:

"The importance of a local network. I asked for help to friends who spent a lot of their time to help me with the website and the Facebook page" (Cristiani, E., Personal Communication, October 17, 2012).

Often crowdworkers need to acquire new skills and upgrade their product or service. Alongside classical methods (training courses, mentoring) the "know-how" culture is now emerging. Self learning methods guarantee more independence and make it possible to experiment following your own path. Many prefer the "trial and error" mode of learning and internet is a valuable reservoir from which to draw on-line information and tutorials for replicating and experimenting on your own.

I don't think I have to go to workshops or courses of illustration to improve my job but I think I have to do more experiments, it's like a trial, I say "ok it works, or it doesn't work", I find that producing and drawing every day and trying different things I can learn more than if I would go to any course. (Lapo., Personal Communication, May 3, 2013)

"I watch video tutorials online. You see how people do things, then you try to do the same, and try, and try until it works" (Oliveira, F., Personal Communication, May 3, 2012).

Often to be a founder or member of a local cultural association during college is a way to learn new skills by doing:

"Learning by doing: organizing cultural music and cinema festivals with friends, so we had to do much things by ourselves so we learned by doing" (Roeder, M., Personal Communication, June 10, 2017).

This type of learning by trial and error is quite rare in schools. Young professionals coming from well-equipped technical schools are not used to work with what is available. Also, in a university environment the trial and error method is quite rare, therefore this is a recurrent characteristic of most community based shared spaces.

This openness to new forms of education allows to experiment new forms of educational organizations too, that can help to make a more fluid transition from school to work:

One goal of our organization is to become a bridge between academic institutions, universities, and the civil society…A present project: we're testing a 3D scanner for the textile industry and working on new ways of education, and a spectroscope made with the 3D printer. (Parade, M., Personal Communication, July 15, 2013)

The financial crisis for the school is the occasion for a transformation of society. A society that had still the idea to wait for the work that was provided by third parties. We as a school have the problem, we put it as a point of reflection together with the boys. We are focusing on a fundamental educational goal: every boy has to point to himself, believe in his own possibilities, no longer have to delegate to others or expect the job from above. (Pongitore, N., Personal Communication, January 27, 2013)

Now, politics is no longer there, and the guy who does not want to leave our region has to collect all the energy, convey to the idea they feel more valid and pursue it with tenacity. This is precisely the first quality that we must teach: tenacity, not to give up. (Iannini, A., Personal Communication, January 27, 2013)

Crowdworkers have made very specific choices about what to do in their life, but their experiences may give some answers to the doubts that today's students are forced to cope with, such as the need to invent their own future, to decide where and in what to invest their time and their own energies:

Somehow we make culture, with the work that we have created and every time we teach something to people who work with us. Just doing that thing, is already doing culture, because people who work with us have learned a craft and already that for us is culture. (Morosi, M., Personal Communication, October 10, 2012)

We must consider that the working world has changed, it is not the same as it was once, so when we say that young people invent jobs, it should be understood without diminishing them because there is actually no other possibility, so if you want to do something, you have to create it, if you can then succeed it is even better. (Sgarbi, F., Personal Communication, March 5, 2018)

The message that the Co-manufacturing Space manages to convey in a concrete way is the need to explore new paths and to build opportunities that are personal, but can be realized only together with others.

"It's great, because everyone helps each other" (Liceo Erasmo da Rotterdam Student, Personal Communication, March 4, 20113).

I call them all artisans, in the good sense of the word, because the craftsman is the one who creates, who creates something that he knows how to do. And then when you see your project realized you are very happy and also the others even if they work in different sectors, they help you, support you, you do not feel alone. (Sgarbi, F., Personal Communication, March 5, 2018)

THE INVISIBLE FACTORY

The Invisible Factory is a structured way to collaborate and to create joined projects that take form in Co-manufacturing Spaces.

What Is the Invisible Factory

In most of the shared workplaces visited during the research, it quite often happened that one of the companies received a request, which needed multiple skills to be satisfied, and thus it created the right team for that specific project.

The company that received the request usually created the team joining other companies that were part of the space, building a situational value chain. The result is that the companies could act not only as a consortium of enterprises, but as one single entity and this unity is so beneficial that usually it is even not necessary to make a distinction between the different companies:

Oficina Colectiva is not a Company, it's a place. But when we work together on the same projects, we don't even have to separate the teams, as we're from the same space, the same environment. We don't have to say, "these are the guys from Toyno, these are the guys from Nambana." (Quinta, R., Personal Communication, May 5, 2013)

When this happens, the team operates as if it was a structured company with separate functions, where well-established relationships, mutual professional understanding, and the habit to work together creates fluidity in the exchange of internal information and in the relation with customers.

During the Bigmagma experiment this also happened several times . For example, there was the request for the packaging of a new bottle of fine wine. The graphic designer made the label, the designer developed the bottle holder, the architect designed a stand for a wine fair, the artisan built the shelves and other objects that were needed.

In another instance, some members of Bigmagma worked on a common project called Grinding Project, an upcycling project aimed to develop, promote and distribute design objects made out of used parts of engines.

The most interesting outcome was that, working this way, from the outside Bigmagma members were not seen as independent companies working together, but they were perceived as divisions of a single, huge company.

Customers reached Bigmagma through one of its members, and then this one involved others in order to set up a specialized team made of single companies, who knew each other very well and could work together in a simple, effective way. This way, Bigmagma could be a clearly identifiable landmark, and at the same time a polymorphic collective entity. This requires a high level of trust and organization between members, but makes it possible for each company to act as an independent economic agent, and also, when needed, to be part of entrepreneurial projects between equals bringing to a specific project only their stronger skills.

We present ourselves as a Project, but if we receive a very specific architectural request we present ourselves as architects. The events and exhibitions are always made under the name of Oficina Colectiva. (Carvalho. F., Personal Communication, May 5, 2013)

As this is not an *a priori* defined and static structure, Companies can act as individual subjects, but also realize a unitary value chain that can deliver a value of output that is greater than the sum of the value of the individual entities that are part of it.

Focardi & Salati named this phenomenon "Invisible Factory".

The Invisible Factory is a system of interconnections between several economical agents within a shared workspace.

These connections are in a sleeping state, "invisible", until some node of the system activates the "Factory" in order to satisfy a need that comes either from an external client, or an internal project.

The Invisible Factory concept refers to the framework of sharing economy, based on access to means of production without having the possession, and this means that artisans and professionals can add and mix their skills and equipments in order to set up a bigger value chain, in order to face complex projects together.

How the Invisible Factory Is Activated

As previously stated, in a shared workplace it is possible to find companies that operate within completely different sectors; their activities are contiguous and this, coupled with the general attitude to collaboration, often leads to joints projects.

People like to share their knowledge, so we do a lot of workshops where members present themselves, and maybe someone says "oh, you're an expert in social media, could you help me with this project?" and they start to collaborate. ... Then there's a third kind of collaboration when someone has just a really brilliant idea and just writes it on a board asking who is interested in it. It's very organic, we don't really have rules and instructions, people just grow as they like, and this was pretty effective until now. (Tham, C., Personal Communication, September 18, 2013)

The first step to make the Invisible Factory alive is to build a structure that allows crowdworkers to establish a relationship of confidence with each other. Common events not related to working purposes are a good starting point for getting to know other approaches to life and other mindsets.

We have common areas where there is the lounge, where we eat, the space where we play ping pong, where we do yoga. But basically then each one has its own working space. Sandra and me we are in the steering committee of the cultural association that manages Balnerario and we handle with the organization of the common activities. But we have a very busy work with our studio and there is no person who is devoted to just run the association. So it's a very free thing, anyone can arrange or contribute to the realization of joint activities. (Brazao, R., Personal Communication, May 5, 2013)

Hanging around the space on a regular basis is the first essential element to activate the Invisible Factory. Why is a physical space, where people go and work and meet, so important in the era of technocracy? Because of two simple reasons: diversity and coincidence.

Diversity of People

Crowdworkers can work in completely different fields and can have such different backgrounds that, as a consequence, they have completely different tastes and approaches about anything, from music to fashion, from religion to family. As said at the beginning of this book, this complete diversity is one funding element for Co-manufacturing Spaces. As a consequence, inside the physical space one can meet someone that they otherwise would never have the chance to encounter in their whole life. This complete diversity, as we have seen, is so beneficial for these entrepreneurs that when they perceive how positively it is affecting their business, this becomes one of the reasons that make it worthwhile to go out from home, to pay for a space in a shared workspace even for those that don't strictly need a space for running their activity.

I really like the fact that there are companies that are totally different from us but that's just what stimulates, the fact they are different: first of all you become interested in new things, you learn new worlds. (Sgarbi, F., Personal Communication, March 5, 2018)

The physical space is a key point to activate the Invisible Factory, as professionals can meet each other by coincidence. It is not required to arrange a meeting, to find time for an appointment to talk about something. Encounters, repeated over time, happen by chance. Exchanges become faster.

"Rosenwerk is a concentration point. The opportunity of coincidence is higher here, more people are moving around, and doing some stuff and there is more opportunity to meet other interesting people" (Roeder. M., Personal Communication, June 10, 2017).

Joint projects, then, can start randomly but they can also be planned or grow around a chance sometimes offered by the managing entity of the space. This is the case of an opportunity coming from outside that explicitly requires a pool of multiple actors, such as to apply to a public tender, a Call for Projects, a common event or a common faire.

The Invisible Factory activation can take place when a request comes to one of the members of the space, who responds to the request and activates the invisible factory involving others which have the skills needed to respond to that particular request.

In other cases, there can be an explicit request from a potential customer sent directly to the shared workspace as a whole. This happens especially in the case of spaces well rooted in their surroundings, when Customers with different and multiple needs try to find a unitary solution in a single space that is a clearly identifiable landmark, with a territorial dimension.

There are situations where someone from the outside says, "I have a product, I just came up with this new wine, but I need a team, can you put a team together? So we say, yes, what do you need? I need a photographer, a graphic designer, a marketing person, and we have all of that here, in MOB. So we build a team and sell it as a package. (Tham, C., Personal Communication, September 10, 2013)

For bigger and more complex projects, one of the companies involved becomes a contractor and manages, on behalf of the whole group, the relationship with the customer.

In other situations the Invisible Factory is activated directly by companies that are part of the space, that start joined projects to propose new ideas on the market.

To build the microfiliera we started out from market needs. We collected Evra's solicitations and then we later identified the farmer entrepreneurs fitting better to this project. In this way we have created a bridge between business and market. (Gerbino, D., Personal Communication, January 24, 2013)

A Light and Flexible Work Organization

Compared to other modalities, this way of organizing the work has visible advantages in terms of efficiency.

The Invisible Factory is a latent structure that is activated only when needed and is reconfigured each time. This allows to better optimize the skills of each company and to offer a complete package, homogeneous and functional, perfectly fitting the customer's needs.

This has another immediate advantage, given by the fact that the team can be reconfigured anytime in a different way. Even though the departments of this "factory" are not always the same, all participants to the project know each other personally and professionally.

In a shared environment, every member is aware of skills, possible roles, specialties, strengths and weaknesses of the others. When answering to a client request, a member can rely on a full system of available skills and competencies.

Within this structure, each artisan or freelance, can be both an independent economic agent, which bases its activities exclusively on itself in every way, but becomes part of an entrepreneurial project between equals bringing to a specific project only its stronger skills and the organization of economic and operational relationships is based on bilateral or multilateral agreements without a top-down management.

Starting from skills when organizing the production process allows to organize the work easily and with a light structure, without those fixed costs typical of a standard factory.

In case the Invisible Factory is activated to explore an idea for a new product or service, the Invisible Factory becomes a good tool for answering to the reduction of the life span of ideas and the life cycle of a product:

Many people talk about brilliant ideas every day, all the time, hundreds of ideas. I don't think an idea is a personal thing, an idea comes at the same time all over the world. We just want to go ahead and make it happen rather than talk about it. The important thing is to realize your idea and to make it happen. Or at least try: take a risk, do something different, doesn't matter if it doesn't work. (Nichols, A., Personal Communication, February 27, 2013)

The opportunity to approach the market together allows small economic agents to access projects they would not be able to handle alone, and to tackle big projects with the benefits of a multiplicity of ideas and views along with the benefits of an already structured coordination and a collaborative habit that makes the workflow fluid.

The invisible factory also protects the different stakeholders. Even when working as a unitary subject, the Invisible Factory team, each company remains separate and independent. The team is not exposed to banks as a whole and if a member comes into financial crisis, it would not put the project in danger. It follows that a single component of the system may have problems, but the system as a whole is protected. This is a form of protection for the team and its components, but it also protects customers, and suppliers.

Also investments are widespread and even when dealing with large communal projects, the total investment is the sum of the investments of individual subjects. These investments first lead to local consequences, related to the professional daily life of the individual companies, and then follows a common effect that is a system effect. The result is that the overall effect you get is more than the sum of the individual effects as every investment is made once, but has two effects, a local and a systemic one.

The Invisible Factory is a structured way to organize collaborations within companies based on Sharing Economy ideas, but shifting from more traditional organizational system to the Invisible Factory is not easy and an open mind is needed.

We also support people with business ideas, but only businesses that want to change the way to do business. Not based on concurrence but based on collaboration. How can you base business on collaboration is a big question? This place is the place where to make the experience to answer to this question. (Roeder, M., Personal Communication, June 10, 2017)

Sharing and collaborating is not so simple when it comes to interactions between companies, artisans or other kinds of economical agents, and is important that every shared workspace, as a community, can give guidance, training and tools to its members, to help companies in their effort to share and collaborate, to avoid misunderstandings and solve issues while managing shared projects.

CONCLUSION

The extensive research performed by the Authors within different shared workspaces had different outputs. The main topic is how the meaning of work is changing according to new needs and a new self-representation of independent workers, and how new kind of workspaces are hosting and fostering this change. An important point that emerged is that these environments are socio-economic systems suitable to activate collaboration dynamics and to foster exchanges practices between members, that become a different way to start, grow, manage and organize companies, also if shared spaces are not a suitable solution for everyone.

The first phase for a crowdworker is the most difficult and delicate. There are those who expect a service and there are those who are not used to a non-vertical working environment.

At first, when I was here, I was very timid and did not know if I could really be myself. I was shaken to work here together with others because people at first should try to figure out where they can be, how much they can open up to others, then, understand this, relax. (Lapo, Personal Communication, May 3, 2013)

Another limit, albeit in the opposite direction, is the difficulty of concentrating on one's own business at times when there are proposals, opportunities and external stimuli.

This is sometimes not a good thing, in fact, there are days when I come here I can really do so little because it's all too much fun. There are times when we talk so much between us or play table tennis, last month I had to work two days from home because I could not get involved in work because so many amazing things were happening. (Lapo, Personal Communication, May 3, 2013)

The relational structure leads to very close relationships between crowdworkers, which often lead to friendly, interpersonal relational structures, with the following advantages and disadvantages:

It's a model that has already proven to be very efficient. First of all, it is very exciting to work with friends. It gives you many benefits, you can have fun working together and many people do not have the opportunity to work with their friends. It's certainly very important to define working relationships, but if you are aware of what pitfalls you can conceal and how can they be avoided, the benefits of working together become very clear. (Parra-Mussel, A., Personal Communication, February 27, 2013)

Another important factor is to identify and maintain a balanced breakdown of roles and tasks such as welcoming new members, organizing public opening, routine maintenance or managing utilities.

They actually feel like a family here. So when they come again here after two years, they don't feel they are going to a random place but to going back to visit friends, also. This is the main think about the community. The step forward a community, is family. We do a lot of things to make a community really solid. (Toriz Morales, M., Personal Communication, August 24, 2016)

Whether there is a formal and informal management body, this is a matter that, if not dealt with caution, can put the entire system in a crisis.

"A negative aspect is tied to all the spaces where many heads live together and a synthesis must be found to make the different heads pursuit a common goal" (Cristiani, E., Personal Communication, October 17, 2012).

Being able to build a common vision and taking the time to write a manifesto in a participatory way is a great solution that allows to set a common direction.

Co-manufacturing spaces are cultural landmarks that need the support of public institutions. But for public institutions it is not easy or even possible to support private companies. Another challenge is to make it easier for the public sector to collaborate and embark on joint project together with shared workspaces.

Finally, an important aspect is that a Co-manufacturing Space is largely self-organized and that relationships are peer-sized. There are no support structures from above, there is only a peer community that includes some forms of mutual help, but it's also an environment that brings new members to face very quickly and violently with their own mistakes.

REFERENCES

Chesbrough, H. (2003). *Open Innovation: The New Imperative for Creating and Profiting from Technology*. Boston: Harvard Business School Press.

Focardi, G., & Salati, L. (2015). A New Approach to Knowledge Sharing, the Multifactory Model. In O. Teràn & J. Aguilar (Eds.), *Social Benefits of Freely Accessible Technologies and Knowledge Resources*. Hershey, PA: IGI Global. doi:10.4018/978-1-4666-8336-5.ch009

Focardi, G., & Salati, L. (2016). Multifactory: An Emerging Environment for a New Entrepreneurship. In N. Baporikar (Ed.), *Handbook of Research on Entrepreneurship in the Contemporary Knowledge-Based Global Economy* (pp. 115–129). Hershey, PA: IGI Global. doi:10.4018/978-1-4666-8798-1.ch006

Salati, L., & Focardi, G. (2014, June 13). *Multifactory. The Invisible Factory* [Video file]. Retrieved from https://www.youtube.com/watch?v=n6ZnhC7B9iE

Srnicek, N. (2016). *Platform Capitalism.* Polity Books.

Chapter 3
Crowdworkers as a New Social Class

ABSTRACT

Crowdworkers are a social class which takes form from a common social and economic situation. People in a similar situation organize themselves as a whole, in order to increase their chances to improve their lives. These people were similar among different countries, and expressed similar concepts, values, lifestyles. They organized their professional activities in a similar way, had similar goals, and in the workplaces the overall organization of the relationships between members of the co-manufacturing spaces reflected these similarities. This chapter underlines some important aspects that lead to the idea of crowdworkers as a social class. Crowdworkers are economic agents and the change in society they bring is mainly recognizable at the level of economic dynamics. That said, economic dynamics come with social ones and can have a huge social impact. Although it is never easy to define how social and economic dynamics interact and influence each other, under a general point of view the authors agree with the conclusions of Federico Caffè, who believes that in society the driving force is social, and economic dynamics follow the social ones.

DOI: 10.4018/978-1-5225-7089-9.ch003

NEW WAY TO ORGANIZE WORK AND PRIVATE LIFE

As Amoroso (2016) underlines, "Caffè was convinced that the economy is an important tool for the wellbeing of people. But of people within a system of emotional and social relationships from which they derive their vital lymph to which their affection and commitment must correspond."

Co-manufacturing spaces are places where these emotional and social relationships occur between crowdworkers and are accumulation points that can foster meetings between similar people which have some goal in common.

You are willing to get involved, because no one would get involved in something in which he does not believe to the very end. All those who are here in R84 decided to transfer an activity to a new place, and this is not such an immediate decision, so I'm convinced that no one would take it without thinking about it. So if all of us were willing to do so, it means that we all believe in it. And this will allow a great bond and create a great strength between us. (Scattolini, C., Personal Communication, March 5, 2018)

Crowdworkers, as diverse as their ages, paths, experiences, and specific professions might be expresse a unified and recognizable model of doing business and living life. Priority, lifestyle, values, motivation, choices of life, income, self-perception of their social status.

They are a source of inspiration for young people and are a set in many ways blurred, but overall very recognizable.

This element of self-representation, the sharing of values and experiences as a result of professional experience, the same social framing, are elements that historically define a social class and of this there is awareness, at least in part.

We are a mix of working class and middle class. The class system is changing. We are all working class as we do things but we absolutely are entrepreneurs. I've not seen this particular model we are building BB in operation everywhere else in the world, so I think this is entrepreneurial. We are business men, but not exactly like most business men, as they are supposed to have profit in mind, whereas our is not for profit. (Parra-Mussel, A., Personal Communication, February 27, 2013)

Crowdworkers experiment a new way to manage their professional connections. Relationships between different companies, freelancers and other economic agents within Co-manufacturing Spaces are economic ones. Once established that, not one of them works with others solely on the basis of economic/business considerations but also on the basis of trust and personal empathy.

It is a two-stage process where empathy is the first filter and economic considerations the second one. Therefore, each Company chooses whom to relate to from a value-based standpoint and then cooperates only with those that it considers economically advantageous.

How to identify and manage social dynamics, especially among players with such diverse origins and experiences, is something that the group of peers learns over time.

In proposing the micro supply chain, the difficulty at the initial stage was mistrust, because producing and selling a product is not something easy, given for granted. And in making the farmers meet with the companies, in this phase we have taken on the role of the guarantor. All this work has served to make the various subjects talk to each other and now it's all easier. (Gerbino, D., Personal Communication, January 24, 2013)

Therefore, economic and social dynamics are not only intertwined but they are interdependent too, and this balance between different levels concretizes an environment that combines economic sustainability needs with participation, proactivity, sharing, opening and exchanging.

This balance is one of the cornerstones on which the concept of social enterprise should be based, and this is the direction the discussion is going at the European Institutions level. As was apparent in Strasbourg, with the interventions of Barroso and Schultz, during the first conference addressed to social entrepreneurs: *"Entrepreneurs sociaux: Prenez la parole!"* January 2014.

The concept of social enterprise is evolving, moving from a classical understanding of an enterprise which only has to deal with social issues (such as childcare, teenagers, elderly people, patient care, sheltered groups in general) to a wider definition, which, independently from the sector of affiliation and from the product or service offered, the company in pursuing its own normal business also achieves social purposes.

In the case of a Tool Library, for example, the business model is clearly going this way:

People take the classes for fun and entertainment, if they have extra income to pay for these classes that's awesome. And we use that to be able to provide tools to people at incredibly affordable price. A smart organization can make it's own money without having to ask for everything. And when you can apply the design thinking towards generating revenue and building a smart organization they can have a very strong social mission and be a successful kind of like a social enterprise in a city environment, that's the most important thing. It is different from a capitalist model but we still try to make money where we can. (Shar, J., Personal Communication, July 23, 2016)

Therefore, crowdworkers offer peculiarities that are not apart from other society phenomena but are elements of socio-economic innovation that fit into a wider debate and answer questions already fitting into the macro-economic system.

Up until a few years ago, there were various contexts in which people did things for free to support a specific community they belonged to. Just think of a parish, a voluntary association or a political party. This activity was perceived as a contribution to a common cause. There was therefore, at the base, an attachment to a common value structure and the desire to do something concrete to help the micro-community one personally felt to belong to. This way of behaving was normal also within various working environments: between waged workers, but also craftsmen working in the same neighbourhood, or local shopkeepers.

This mutual support gradually disappeared. The general feeling is that cities have become more and more individualistic communities where many people live similar lives and have similar experiences; they are often frustrated in precarious working conditions, but they keep their status for themselves, and no longer feel the desire to offer their time and their presence for free to others.

This mechanism puts the individual in the foreground, but the community goes in the background and, in the end, this dries up local communities and the territory they belong to. This is a key point in the writings of Federico Caffe', and Amoroso (2016) reminds that "studying the systems of social and production organization that had established itself in Italy in the 1960s and 1980s, the industrial relations systems and the management of the State of Wellbeing - Caffè clearly identified the growing contradictions of a system that increasingly replaced the affective and communitarian values of solidarity with the mercantile ones of "negotiation" and "contract" with the consequences of merchandising and monetizing the social ties."

Crowdworkers, however, reaffirm the value of giving spontaneous help not aimed at immediate economic return, and usually offer each other moral and material support, both for practical aspects and as a form of exchange of professional skills. These are exchanges which do not offer the prospect of an immediate return or a direct personal advantage and follow the simple desire to be helpful.

It is a free exchange based on the sharing of a common condition, on recognizing part of the same project and on the will of being an active part in building a common destiny.

However, this is not a precondition. The reason why someone joins a shared workspace is not because he identifies himself in explicit common values, as could be the case when entering a religious group or a political organization.

Everybody comes in our space. It's king of like a leveller. Everybody can get access to tools. Even people who are incredibly wasteful still would like to save money. But also a lot of people enjoy being a part of a community too. People will start to recognize how much better is to be a part of a community instead of processing things on their own. (Shar, J. Personal Communication, July 23, 2016)

A shared workspace is a community of purpose; it is firstly a place devoted to work, where the common values rise up slowly and autonomously, thanks to a participatory method used for common projects.

In this context, to offer free help becomes a natural way of giving mutual support, which is not a mere do-ut-des exchange nor a sign of moral superiority, but a deep participation in a working and existential condition. It's a system that brings people to make things and to help each other, reminding the vision of Schumacher (1973) in *Small is Beautiful* "dreaming of systems so perfect that no one will need to be good."

Of course, this approach also brings economic advantages, as it becomes a value so incorporated in the companies that it becomes a very strong intangible asset, and this allows to bring to a Co-manufacturing Space also free help and support from outsiders.

It's very difficult to go along and find somebody who says ok, yes, I really really want to give my time and help with my efforts in order for you to get wealthy, as soon as we made it a non for profit organization suddenly the flat doors opened and people come to invest time and ideas to help us create the vision we had in mind and when we removed money as the primary objective of the company suddeny the development possibilities grew enormously. (Parra-Mussel, A., Personal Communication, February 27, 2013)

It is noticeable by everyone, even coming from outside, that co-manufacturing spaces are economic stakeholders, but it is equally clear that the priority in general is not to make profit but to set up a peaceful life and to guarantee themselves a general wellbeing.

We all develop something related to creativity, we like to be together we do many things together I do not know how to organize exhibitions, cultural events and we all have our active professions that we develop independently of this. The goal is to create a beautiful atmosphere here and that people feel good, feel like in a small family, but this is something that comes naturally, which can not be decided a priori and it is not even something you see immediately when someone comes in again because it is related to personality. (Brazao, R., Personal Communication, May 5, 2013)

WORK CULTURE

The interviews show that in general crowdworkers have a completely new set of meanings of what work is and what one should work for.

Previous generation had this conception that you study hard, so you can go to a good college, you get a job, a good job in a big company, then you stay there for 20/30 years, you earn money, so that you can buy a house, you know, get married, have kids, then you have your formula of happiness. Nowadays a job is what you invent yourself, it's no longer just about getting the education that you need and applying your education to somewhat else, to a corporation or an institution that needs your skills. Nowadays a job is what you can invent, you can be whatever you wanna be. Tham, C. (2013, September 18). Personal communication.

Shared spaces are closely connected with the neighbourhood in which they are located in many different ways.

The first one is if some part of the space is facing directly to the street: they are not shops, neither factories, there are some products on display but also some people working in front of a laptop or making physical objects with a machine like a 3D printer or a drill. From a public perspective, these spaces are something hybrid, strange, that attracts curiosity.

"To be on the ground floor is a way to invite people to come in. They come, they ask about the space and the project." Carvalho. F. (2013, May 5). Personal communication.

Then there are many other deeper connections with the surroundings that lead to some sort of interactions with other local actors. In certain cases these connections are a planned strategy by some companies specialized in the cultural and social innovation field.

In other cases it is just the general openness of the space that leads to positive exchanges with local cultural public institutions as, for instance, high schools.

For its inner characteristics, a co-manufacturing space opens up the possibility of offering guided tours, stages, internships, and collaborations in general with the local community.

A positive result of this openness is a direct contact between youngest generations and a working world that is drastically changing.

Kids need to get their hands dirty, practice the activities they are studying. If we first tried to circumvent the problem by simulating the disadvantages, today we go directly to the company" (Iannini, A., Personal Communication, January 27, 2013).

School systems started to rethink their relationship with the labour market when youth unemployment critically arose along 2008. Suddenly, there was a need to find a way to help students make a change of perspective about how to enter the labour market.

Culture is no longer a set of notions, it is to put the kids able to move independently into the world of work and to be citizens of the world. They have to live their lives, not the lives of others, but to live their lives, the boys must have the tools to be competitive in a competitive society. (Pongitore, N., Personal Communication, Januar 27, 2013)

Shared workspaces slowly become landmarks: for instance, some teachers from the Human Sciences High School *Erasmo Da Rotterdam,* based in Sesto San Giovanni, have contacted the Bigmagma multifactory close to their school for their yearly school trip to a popular industrial plant, symbol of the town. In the Milano area, this usually was a guided tour to the headquarters of the *Corriere della Sera,* the national newspaper or to *La Centrale del Latte di Milano*, one of the most popular plants making milk-based products. But in this case, these teachers instead chose to contact Bigmagma and to offer their students a *Mage Tour*.

Part of this unusual factory tour was a videobox where to record the vivid impressions from the students, a way to give them the opportunity to immediately and openly rethink about the experience:

"I'm impressed with creativity, they all have a lot of inventiveness and courage to get in the game, so congratulations" (Liceo Erasmo da Rotterdam student, Personal Communication, March 4, 2013).

"It is thought that children have more imagination than adults and instead seeing these people here, you see that it is something that can be carried forward over time" (Liceo Erasmo da Rotterdam student, Personal Communication, March 4, 2013).

A co-manufacturing space can give the will to be bold, to take a chance and to go for it. To find a way to make it sustainable.

"So many different people have so many things in common, because they all work for a common goal, that is to find a different way to start an activity" (Liceo Erasmo da Rotterdam student, Personal Communication, March 4, 2013).

"It's amazing to see people with the courage to do what they want, to start from your own idea and to keep on going" (Liceo Erasmo da Rotterdam student, Personal Communication, March 4, 2013).

Co-manufacturing spaces spread the culture of making experiments, learning by doing, through a process of trial and error more close to younger generations than to the older ones.

"There are no perfect places, or perfect projects, you simply have to try your own way" (Parade, M., Personal Communication, July 15, 2013).

This approach is closer to the world of younger generations who are increasingly taking a distance from the adults because of a different way to communicate and to live relationships. Crowdworkers are forced to learn and use the same language of the younger generation as they have to deal with the same means of communication used by young people, not because they like it but just simply because Social Networks are the cheapest tool to promote their business. New languages have new grammars and social networks are much faster than other languages so adults have to learn, for example, how to deal with imperfection, they cannot wait to go out with a video or post until it is perfect, they need to learn how to deal with imperfection. A shared workspace is much closer to a younger world than other contexts. Everybody is forced to listen to other people, to break some mnemonic barriers and to enter into dialogue and exchange with them, to be open to learn from them.

This openness to giving more value to their experiences is appealing for young people, who are curious enough to enter in touch with a co-manufacturing space, and experiment a feasible new path offered by the current flow of new works and business opportunities. They can see what's new, how to organize life under a new perspective and they can decide what fits them and what doesn't.

"You have to believe, you have to do it. This experience teaches you to believe in small things, you do not need to have to have big resources, just [be] inventive" (Liceo Erasmo da Rotterdam student, Personal Communication, March 4, 2013).

"I do not know if I could [do] the same, but I was struck by the courage of these people" (Liceo Erasmo da Rotterdam student, Personal Communication, March 4, 2013).

Courage, inventiveness, to reuse skills and experiences to build up their own work, the ability and the will to try to shape their passion into a profession:

"I would love to become a photographer, though I know it's difficult. What I have learnt today is not a job, it is a philosophy of life, that is to recycle and make weaknesses point of strengths" (Liceo Erasmo da Rotterdam student, Personal Communication, March 4, 2013).

For crowdworkers, the work becomes the way to express themselves and to reach, to some extent, fulfilment in their life.

"Work is the personal fulfilment for everyone, work should be the utmost expression of each person" (Saiu, F., Personal Communication, October 20, 2012).

CROWDWORKERS AS CULTURAL AGENTS

We will re-inspire people to innovate again rather than following set rules. We will allow people to try their ideas and when there is more than one person trying an idea it's much more interesting and we can go a lot further with innovation and new ideas in a surrounding which doesn't restrict it. (Nichols, A., Personal Communication, Febrary 27, 2013)

Crowdworkers as a whole are aware to be agents of a cultural change even if most of them are not cultural managers.

Co-manufacturing spaces are also home to some kind of companies dealing with the cultural industry, as common heritage and museums innovators or cultural event planners. They can also welcome associations run or initiated by cultural and social managers, as in the case of *fabLabs* or *Repair Cafe's*. In these cases, the culture of sharing, open knowledge and open source is something that is well identified and visible.

But contrary to what one may think it is not the availability of new technological equipment (3D printers, laser cutting machines) to foster a knowledge-sharing culture:

It's not so important to have very expensive tools, as you can't use these tools if you don't understand how to operate them. This is to us very important: we try to make a new way to find new technologies and open these technologies to other people, and we're not alone in doing this. (Parade, M., Personal Communication, July 15, 2013)

This free knowledge exchange works pretty well in cases where the formal or informal coordination of the co-manufacturing space can support and promote the cultural activities carried out by the FabLab community, such as open days where to meet each other and talk about technology-based projects or events to bring this culture and approach to traditional public spaces.

Since 2012 MachBar, the Fablab initiated by Mario Parade inside FreiLand, a Cultural shared space based in Potsdam (DE), runs a monthly *Repair Cafe'* at the Central Library of the town. This has been a minor action with a huge following. To bring power tools to a public library, to be open to help and teach to normal citizens how to repair something has an immense cultural value. A *Repair Cafe'* is an on-going learning process that makes a bridge between FreiLand and the outside world. It brings to light, available and understandable to everybody, some of the common values and beliefs that shape the microcosm of a co-manufacturing space.

As stated in the international repair café manifesto (2016) "Lots of people have forgotten that they can repair things themselves or they no longer know how. Knowing how to make repairs is a skill quickly lost. Society doesn't always show much appreciation for the people who still have this practical knowledge, and against their will they are often left standing on the side-lines. Their experience is never used, or hardly ever.

The Repair Café changes all that! People who might otherwise be side-lined are getting involved again. Valuable practical knowledge is getting passed on. Things are being used for longer and don't have to be thrown away. This reduces the volume of raw materials and energy needed to make new products. It cuts CO2 emissions, for example, because manufacturing new products and recycling old ones causes CO2 to be released.

The Repair Café teaches people to see their possessions in a new light. And, once again, to appreciate their value. The *Repair Café* helps change people's mindset. This is essential to kindle people's enthusiasm for a sustainable society.

But most of all, the *Repair Café* just wants to show how much fun repairing things can be, and how easy it often is."

People who are able to fix things, to open a monitor and understand how it works are becoming such a minority that even inside the co-manufacturing space these positive and unexpected encounters are well visible.

The close proximity between fabLab members (mostly nerds, hackers, drones and 3D printers enthusiasts) and employees or owners of small and medium companies is quite unusual.

Christa Müller (2013) highlights how "A new generation of do-it-yourself activists is using the post-Fordist city as a laboratory for social, political, ecological, and aesthetic experimentation. Whether in a common garden or in a FabLab, in open workshops or trade events everywhere, such protagonists question the relationship between consumption and production and expose the problems of the commodity character of things and the knowledge contained in these."

Outside this environment they would hardly cross each other's path, but as far as they meet in these experimental environments the potential transfer of a latent culture allows all actors to benefit from it.

Knowledge sharing is the founding element of a FabLab or an Hackerspace. This concept and the philosophy behind it can be easily understood and accepted by crowdworkers, who are more open than traditional entrepreneurs to any kind of new ideas, no matter their business field, and are also willing to break the barrier between technology and the world of doing practical things:

Most people think that design thinking is all about thinking, but in reality what makes it different from other processes is the act of doing, using your hands, prototype really quickly, and see if it works. If it doesn't work, rebuild it, and then do it again, do it again, do it again. (Quinta, R., Personal Communication, May 5, 2013)

Social or Cultural Promotion Associations as Art Collectives can take a lead in promoting this cultural values because they have the diffusion of culture as a statutory goal.

In general, crowdworkers have a medium-high cultural level and are sensitive to cultural issues, in different forms and declinations, and not necessarily within the working environment.

"We do not make culture, we do business even though we are engaged in culture outside the working environment" (Bocola, M., Personal Communication, October 1, 2012).

"I had to travel, so I had to compare it with other cultures, so for me culture is to be open, to accept what I don't like. You have to be able to always find bridges with something you don't like" (Saiu, F., Personal Communication, October 20, 2012).

What emerges is that entrepreneurs who practiced more than one profession, who are used to rethink themselves again and again, who learned the essence and positive aspect of skills reuse, have a greater understanding of the value of the knowledge sharing:

"Culture is the dissemination of knowledge to as many people as possible" (Colombo, S., Personal Communication, October 12, 2012).

Culture is a set of values that can capitalize on the knowledge that comes from the past, for example, ceramics is an activity that has a history and the attempt is to carry it forward and transfer it to others, which is why I started doing the courses, to transfer this knowledge. (Cristiani, E., Personal Communication, October 17, 2012)

But at the same time generally they are aware that knowledge sharing is something unconnected with the everyday entrepreneurial activity, it can be a stimulus but it is not a primary goal.

"I would never have an ethical role, we do not have an ethical role, but simply with our daily choices we can help spread some ideas, such as the culture of reuse and reuse" (Colombo, S., Personal Communication, October 12, 2012).

It is a very, very nice ceramic, and without wanting to define art I hope to do things that are beautiful, give the pleasure of being owned and used, maybe not the ultimate art but a concept of beautiful yes, that I would like to be able to carryforward. (Cristiani, E., Personal Communication, October 17, 2012)

"Even if we do not see the Mage as a place for culture, we were looking for a place to produce objects and grow our business in an environment similar to us, so sensitive to cultural issues" (Bocola, M., Personal Communication, October 1, 2012).

"I make culture because I find that in retrieving things that people are not using any longer, transforming their function, putting them in circulation that is a very important cultural action" (Merli, O., Personal Communication, November 2, 2012).

Crowdworkers are focused on their products or services and don't pretend to have a cultural impact as entrepreneurs. They may know that there is a cultural value in what they do, but their aim is not to be recognized as change makers.

At the same time, co-manufacturing spaces become open to external visitors on many different occasions: *Open Days* are a very common practice in shared workspaces. An *Open Day* is when crowdworkers can show the excellence and peculiarity of their products or services, opening for one day their workspaces, offices, ateliers or studios.

Also workshops and courses held or organized by some crowdworkers are a way to let outsiders get in touch with the space and the surrounding community.

Therefore co-manufacturing spaces rather than being places of culture in strict terms are environments where to develop and investigate concepts related to the work culture.

NEW MODELS FOR NEW ISSUES: HOW CROWDWORKERS MANAGE JOB INSECURITY

In the past, at least in Italy, having an open-ended employment contract was seen as a life guarantee. People could plan their lives, buy a house and be more or less secure that they could always rely on said income. As Stefano Mancini says at the beginning of the documentary *Multifactory. The invisible factory (2014)* "This is something true. You can be able to manage your life, buy a house, build a family if you have an open-ended contract".

The current trend, however, clearly goes in another direction. Stable contracts are becoming more and more rare and even those who have an open-ended contract must take into account during their professional life such events as dismissals or definitive closure of the company they work for.

As flexibility becomes a relevant asset for companies, the counterpart for workers is job insecurity, that becomes more and more a structural constant for many workers.

Many researchers tried to define how the change resulting from job flexibility impacts society and social classes, and have often identified a new social class in precarious workers.

Indeed, job insecurity is not simply a condition related to the type of contract a worker can obtain (as the use of short and very short-term contracts), but includes different elements which are linked to each other and put the worker in a weak position. This concatenation of factors isolates the worker from his peers, puts him in direct confrontation with his employer and denies a collective safeguard and redress. In the last ten years precarious workers tried to take some actions together, like the activities performed by the feminine group of "Serpica Nara" in Milan, sometimes trying to be recognized as a new social class, as for instance the movement of "San Precario" did in Italy. And some researchers covered this rising social movement as a first answer to the change in the economic system. Guy Standing in his book *The Precariat: The New Dangerous Class* (2011), blames globalisation for having plunged more and more people into the precariat, which he analyses as a new emerging social class. According to Standing, the precariat is not only suffering from job insecurity but also identity insecurity and lack of time control, not least due to workfare social policies. He proposed basic income as a solution for addressing the problem. Guy Standing states that "is a unique social class truly transformative, a class that when it will becomes strong enough will find the conditions to abolish herself."

So, precariousness is a condition suffered and lived as a negative one and precarious workers aim to stabilize their condition and stop being precarious. Standing underlines that for many precarious workers, "precarious" is an adjective, is not a substantial personal characteristic. It is temporary. Precariat is seen as a transitional state and a personal condition of single workers, and this does not help in finding common values, common issues and a common response to these ones.

Therefore, for some workers, precariousness is becoming a definitive condition but in this case it seems that the theory of Underclasses from Charles Murray (1996) is a better fit, as precarious workers are located at the lower level of society in terms of power and income, and often social status, and they see their situation as a result of events suffered.

Crowdworkers, on the other hand, start from the need for flexibility, and re-elaborate it in a positive way.

Crowdworkers are the social Class of "Stable Flexibility". Precariat is the personal weakness of single workers in a system that was designed to link job stability to personal stability and stability of industries. When companies start to loose their stability, they charge the cost to the weakest part, waged workers. Crowdworkers are an emerging property of a new economic system. They are a new social class that emerge from a changed society. In the new society, companies can't be secure. Industries are not secure. The need for change is continuous and crowdworkers are just prepared to this scenario. Their personal stability is not linked to the stability of the institution they work for, but is linked to their ability to evolve and adapt. It's a different professional skill. In this manner, they can be "stable", that means establish a family, run a business, buy a house, while at the same time being flexible, that means changing while the labour market changes.

An interesting fact to be mentioned recurring in multifactories is the high rate of female presence.

"I also thought that maybe the reason why in R84 there are so many women is because in the traditional world of work we have less chances and therefore we are all willing to engage to do what we really want" (Scattolini, C., Personal Communication, March 5, 2018).

In this new context insecurity is turned upside down and this flexible condition becomes an opportunity to foster innovation, to have new visions and to build new values.

The profession of a crowdworker can suddenly change and can become something new over and over.

A crowdworker can perform multiple professions at the same time and this increases his skills. After a while, this continuous change actually becomes something necessary.

To always find new challenges, never giving something for granted, makes crowdworkers working life more fascinating and not boring at all.

"One day you make a thing, another day you do something else, but I can use the same stuff I learnt at the University to have a different perspective on what I'm doing" (Oliveira, F., Personal Communication, May 2, 2013).

SELF RECOGNITION AND COMMON DESTINY

In this chapter values, belief and attitudes of crowdworkers were presented, as the way they manage their lives and their jobs.

All of these aspects together suggest that crowdworkers, in their specificity, recognizability and representativeness, can really constitute a new social class that can create truly collective productive spaces and create something concrete and tangible within a socioeconomic system with which it does not compete or is in contrast.

According to the classical idea of Marx of social class related to access to means of production, crowdworkers can be seen as a new social class, composed by people who share means of production and have the availability of them as the result of a collective availability. Means of production are individually owned but collectively used, with strong ingroup/outgroup dynamics.

Also according to Max Weber, crowdworkers are a new element of Social Stratification, as they both present peculiar elements regarding their Class and Status.

Class – self employed, not capitalistic, first person involvement, immaterial assets, sharing of knowledge and means of production

Status – medium high level of education, liberal mindset, prevalence of quality of life on money, sort of internationalism.

Also under the perspective of Davis-Moore functionalist theory, crowdworkers are a new social class, as they want to be players of that specific kind of entrepreneurship, and want to fulfil the duties to share and collaborate linked to their status.

Apart from the classical theories of Social Stratification, under a modern point of view the key point is to check whether the members of a potential social class also show feelings of psychological identification and solidarity, share the same conception of society, recognize themselves as part of the same class and are aware to share a common destiny.

As crowdworkers are connected to each other, their professional fate is somehow connected, for better or for worse.

Security for workers who have a parachute when needed. Mutual help gives in a way some kind of self-confidence to all crowdworkers.

It is a path made together with other people, and here in R84 there are several women, will it be a coincidence? And it's beautiful because I think that the people who have made the choice to come and work here must have a lot of

Table 1. Self perception of Crowdworkers from R84 Multifactory

During the first Open Workshop on common values held in R84 Multifactory, entrepreneurs, professionals, artists from the Space gave a clear image of how they self perceive themselves and their workplace: *The place:* *Direct, concrete, honest, welcoming* *A place that is good for people* *A place where you feel well* *Place of contamination* *Place of recirculation of ideas* *Work attitude:* *Not stuck in standard working model* *We do what we want* *We do not care about others opinion* *We develop ideas* *We concretize the ideas* *We connect ideas* *Personal attitude:* *Optimism* *If you can do it, I can do it too* *Opportunity to move as we want* *Freedom* *Innovation* *Ready to catch Latent Opportunity* *Go beyond pessimism* *Honest 'willingness consistency'* *The courage to want it The freedom to do it* *And, above all, the serene and steady belief of being right.*

passion and they believe in what they are doing, each in their own field and scope and this is beautiful because although we work in different environments, when we compare we see that we have a lowest common denominator: We all believe in what we do, this is crucial. (Sgarbi, F., Personal Communication, March 5, 2018)

They all believe but they also know know they have a parachute in case of need, that it is not relying on friendship but on a shared common condition.

Mutual help also works because everyone has specific resources available that they can offer in a barter transaction.

Available resources that can be used as exchange/barter terms by reducing exposure to banks and the risk of stopping the company activity due to lack of financial resources are important shared assets, but are linked to the group size and quality. To foster high skills and valuable equipments is a key point to ensure better future prospects.

Crowdworkers, when seen together as part of a Shared Workspace, are not "individuals" but are part of a definite and recognizable social group on a territory, a unitary subject that can become addressee of direct policies by public institutions, including the local ones.

CROWDWORKERS: WHAT IS NEEDED TO GROW

In general terms, all work spaces need to be identifiable by the general public, to easily understand what they can offer and in what ways can interact customers, suppliers, and institutions that encounter these kinds of environments for the first time.

In the daily life of a shared workspace, however, there is always the need to renew and change, depending on the needs and the availability of individuals, to commit resources and energies in the common project.

We lack a plate for cooking and other small things that we build slowly, there are times when people are more free and they can make a bigger contribution to the group and these are some very beautiful moments. We are a little bit weak on the graphic part, the communicative image of Balnerario, we would have to work more on the facebook page but step by step the visibility is increasing. (Brazao, R., Personal Communication, May 5, 2013)

"More specific equipment and tools" (Oliveira, F., Personal Communication, May 3, 2013).

As far as individual companies are concerned, regardless of individual cases, the business aspect is undoubtedly the one in which most crowdworkers have not developed skills.

(If I were the space manager) I would like to help the entities that are here to overcome their greatest difficulty: to find a market place for their productions. Work together to create a common press office, organize events to attract

customers. The weakness of the entities that are here is not related to lack of imagination, knowledge or inventiveness. The weak point is to reach customers. (Cristiani, E., Personal Communication, October 17, 2012)

On this theme, institutions can do a lot, helping companies not by funding, but by providing opportunities, support and advice to build common commercial structures.

"If there is no entrepreneurship culture, you have to be patient and create the necessary commercial structure to help them to reach the market" (Gerbino, D., Personal Communication, Januar 24, 2013).

The entrepreneur can be able to produce well, but if he does not know how to sell it, then [what]? Giving the opportunity to sell is the prerequisite to be able to produce. Where there are the conditions we create micro-branches, where we cannot, we try to support them in another way. Then when companies grow, to be competitive they have to do many different things, things become more complex. (Gerbino, D., Personal Communication, January 24, 2013)

Even with respect to regulatory or productive issues, a great deal of help from institutions is to provide technical support:

It takes tools: many times companies have problems selling their products. They have no tools to go on the market, such as a warehouse for processing. So you have to rely on others. If we add to this, other difficulties, not just packaging, but also intellectual tools to define prices, discount scales, and a commercial network. The difficulty then translates into the fact that the product is selling to few or badly. Our job is to help these companies, bring them not only to produce well but to commercialize their products. (Gerbino, D., Personal Communication, January 24, 2013)

Schools can also help in other ways to develop social models with interventions at a local level, either by building a different culture of work, or by providing specific skills to support micro-entrepreneurship:

One of the skills we are trying to provide to the students is to be able to understand the world around them. Nowadays, the need is to be able to operate locally not only by staying here and working here, but also by taking care of the environment, because local economic activities, especially on the primary sector, are a form of environmental protection. (Iannini, A., Personal Communication, January 27, 2013)

There is a need for preparation, knowledge, competence and entrepreneurship, with ideas that can be competitive not only here in our small Basilicata, but in Italy, Europe and the world. The winning ideas then go ahead. This has become the goal for us, schoolmen. The crisis can also bring some benefits, if we can turn critical elements into learning processes. (Pongitore, N., Personal Communication, January 27, 2013)

One common need, at least in Italy and Southern European countries is to build bridges between school and enterprises.

When a new paradigm emerges everything changes and nothing can taken for granted, so the first ones to take on this challenge are educational institutions, they need to investigate and help their students to face these new challenges:

The younger generation is definitely frustrated, because they are fully aware of the difficulties of the labor market. We try to offer opportunities, especially in the primary sector, focusing on the valorisation of the specialties of the territory. The primary sector is an industry that can certainly offer job opportunities, if the younger generations become able to build their own job. (Iannini, A., Personal Communication, January 27, 2013)

"The school must start from the needs expressed locally and must provide the skills required" (Pongitore, N., Personal Communication, January 27, 2013)

Every actor must to do his part. The main problem is to be able to guarantee a future to young people and our experience is a model: a school that cares about the future of their kids and manages to accompany them concretely to a job. (Fazzeri, A., Personal Communication, January 27, 2013)

"This experience wants to be a model also for traditional companies, which could incorporate some ideas by our pilot project in their company" (Fazzeri, A., Personal Communication, January 27, 2013).

[the school system / schools] School can no longer just focus on providing theoretical skills, in a classroom, but must be oriented in being a school-lab, it has to be able to offer the skills really needed locally. The goal is to find ways to merge the school system with the production system. (Pongitore, N., Personal Communication, January 27, 2013)

CONCLUSION

Crowdworkers form a Social Class which at the moment is somehow weak, as it has no institutional representatives, but is growing in terms of people and awareness and is more and more linked to a specific class of workplaces.

Waged workers have unions, organizations to counterbalance the system of power, and most of all have factories and offices, places where to work but also where to meet, to exchange ideas and to organize. And if the concept of "factory", or "office" brings to mind very specific environments, they can of course refer to very different situations. A small family run factory is very different from a huge plant owned by a multinational group, but, in general, they both have in common some very typical characteristics, like a pyramidal organization and a hierarchical command structure.

In the same way, Co-manufacturing Spaces where Crowdworkers can establish their activity can be run under different conditions and with different Governance systems, but they are all referring to similar ideas and concepts and they all foster relationships between similar people.

Crowdworkers are similar to each other because they respond in a similar way to new social instances. They form a new social class and it is easy to delineate their characteristic features. It is a social class that is heavily tied to a type of workplace, shared work spaces, just as farmers are tied to the fields, workers are tied to the factory and employees to the office.

People interviewed during the research said that being part of a Shared Workspace provides some form of safety, it gives that kind of certainty, which until now was only granted by long-term contracts: if you are part of a given space you know who you are, no matter what you do.

Being part of a co-manufacturing space is an element of self-representation: individuals or companies that are part of it, feel they belong to a project that is recognized also from the outside, as a whole. This means not only to compromise on concrete issues but also to become part of a shared ethic.

If in the past there were big factories and an individual could declare to be part of a self-representation system, today the fragmentation of work leaves the worker standing alone. The worker inside a Co-manufacturing Space is autonomous but is also part of something bigger, a new kind of factory that creates a new aesthetic of work that fascinates, attracts and engages.

It becomes something relevant, a hub for collecting energy and meanings. You belong to it or you don't.

This, on one hand, simplifies the self-representation of the individual towards the outside world.

REFERENCES

About Repair Café. (2016). Retrieved from https://repaircafe.org/en/about/

Amoroso, B. (2016) *Memorie di un intruso* [Memories of an Intruder]. Rome: Lit Edizioni.

Marx, K. (1984) Capital: A Critique of Political Economy. Lawrence & Wishart.

Murray, C. (1996). *Charles Murray and the Underclass: The Developing Debate*. London, UK: Paperback.

Schumacher, E. F. (1973). Small Is Beautiful: Economics As If People Mattered. New York: Harper & Row.

Standing, G. (2011). *The precariat: The new dangerous class*. London: Bloomsbury Academic. Retrieved from https://www.youtube.com/watch?v=Oo9VaoqMYX8

Weber, M., Baehr, P. R., & Wells, G. (2002). *The Protestant ethic and the "spirit" of capitalism and other writings*. New York: Penguin Group.

Chapter 4
A Conceptual Framework for Co-Manufacturing Within the Economic and Social System

ABSTRACT

This chapter proposes three theoretical reflections on three themes of relevant importance for framing and understanding an economy based on principles of exchange, collaboration, and division. In fact, this seems to disagree with the prevailing orientation of the Western capitalist economy, based on concepts of competition, individualism, asset protection, and hypercompetitiveness. The first reflection, therefore, concerns the relationship between crowdworkers and the capitalist system. The second reflection concerns the type of organization that is typical of crowdworkers, based on the concept of community. The third reflection concerns the classification of all those services mutually provided by the members of the shared work spaces and how such services can be effectively classified as non-economic services of general interest, resolving a common misunderstanding regarding the lawfulness of professional exchanges not regulated by monetary transactions.

ARTISANS AND CROWDWORKERS IN THE CAPITALISTIC SYSTEM

Crowdworkers are Cultural Agents, Social Agents and Economic Agents. They are people who choose and promote a specific lifestyle, based on freedom, self-realization and the predominance of quality of life over money. These people are self-employed and they are reluctant to frame their professional

DOI: 10.4018/978-1-5225-7089-9.ch004

profile in precise patterns. However, as already seen, they propose an idea of entrepreneurship which is strongly linked to the idea of "Making", and they share and evolve many traits of Artisans and an economic model based on the idea of Sharing and collaborating, instead of the mainstream idea of competing. So, talking about emerging economic trends and models can't leave out discussing how they interact and relate with the mainstream economic system, which is Capitalism.

Artisans have existed long before capitalism existed and are often attacked by capitalist companies, which on the other hand can also represent interesting job opportunities for the craftsmen themselves. As crowdworkers incorporate many traits of artisans, it is useful to understand the historical reasons for this ambivalent relationship between artisans and capitalist companies and understand how crowdworkers can effectively capitalize on the experience of artisans and propose an alternative model to the capitalist model, but at the same time can collaborate with capitalist companies.

Capitalistic Companies and Capitalistic Systems

Capitalism is an idea that begins in the 17th century, is shaped in the 19th century and then evolves over a period of two hundred years over dozens of Countries all over the World. It's no surprise that there is no clear and univocal definition of Capitalism and under the same undefined umbrella many different ideas can actually coexist.

This is not the place to depict and compare all the different interpretations of the idea of Capitalism, and it's easy to find out different perspectives just by looking at mainstream thinkers like Marx, Keynes, Schumpeter or Friedman.

Nevertheless, it's useful to briefly underline under which conditions, as intended in this work, it's possible to claim that a Company is a Capitalistic Company, or not.

The capitalistic way to intend a Company, as it evolved in the past century, is based upon division of labor in many small tasks performed by different people and the separation of workers from means of production, which means a separation between the labor force and those who own the means of production.

The Capitalists, who own the "Capital", invest a huge amount of money to buy means of production, and to set and run a Company with the final goal to produce and sell goods and services in order to get a profit and to increase their Capital.

In a capitalist market economy, these Companies are more or less free to decide on their investments and to compete on the market, as the price at which Companies exchange goods and services is mainly the result of a negotiation between the different needs of the economic agents that take part in the transaction.

From this basic point originate many different models of capitalism, as many other elements can interact, like the role of State in controlling and limiting the degree of competition on markets to address other kinds of social needs and issues.

However, apart from any variation and corrective action, in general every capitalist system is based on some main elements: Capitalists, Capital, Private property of Means of Production, Wage Labor, Free Market, Profit, Capital Accumulation, and the way they interact.

Separation Between Labor and Means of Production

A key element in whose interaction is the distance that separates workers from the "Company" as a whole. In a Capitalistic Company, workers, people who are given Wage Labor, don't have a complete view of the Company and can't individually modify the way the Company is run and operates on the market. There is a separation between "Owners" and "Workers". Owners decide where the company has to go and then a management system turns these decisions into practical tasks that Workers, the Labor Force, have to perform.

This separation is true also in the paradoxical situation in which a worker is both an "owner" of the Company that gives him a wage labor. This happens very often, for example both in Public Companies and in Cooperatives. Any worker of a Public Company can own shares of it, so that the same Person can be at the same time a "Shareholder" and a Worker. Even in this case, the basics of the capitalist system don't change as this Person in the role of "Worker" refers to the "Company" as a Whole, doesn't directly own any equipment and has to obey to corporate directives even if he owns some Shares.

In a Capitalist Company there is a defined way to organize work in order to split it in smaller parts and the Capitalists leverage on their workers in order to make a profit.

Artisans: Non-Capitalistic Companies

All of this being said, there are also non-capitalistic ways to organize work and companies and most of them are very common also in capitalistic systems.

For instance, most of self-owned businesses, like Freelancers, Artists, Independent Professionals and Artisans, are actually not organized in a capitalistic way.

Of course, this only refers to the organizational models, how companies are structured. "Capitalism" is a poly semantic word and, for instance, people usually refer to personal behaviors as "Capitalistic" when they are related to a clear focus on money, or banks, or a high rotation rate of personal belongings. Therefore, artisans can be part of the capitalistic culture, refer to themselves as "Capitalists", and integrate in their professional behaviors some capitalistic aspects, as for instance the tendency to money accumulation. In a "Capitalist Society", Artisans get money from capitalist banks, work with and for capitalist companies and are part of the Capitalist System as a whole.

However, an artisan brings to its activity both labor and means of production and there is no separation between owner and worker. An artisan has the complete view on all the phases of work and lacks the typical division of labor of capitalist companies.

When an artisan employs someone, it should seem he acts as a capitalist, but he doesn't.

It's true that in that case the artisan offers wage labor and pays for the work of another person, but the kind of relationship between the employer and the employee is far different from a Capitalist Company.

In the most classic vision of craftsmanship, the artisan usually has one or more apprentices who work for him. They get paid, but what the artisan really gives back is mainly skills. It's a particular kind of relationship and those who work for an artisan want to learn the job, as the apprentice is not a person who is intending to work for someone else for their whole life, but a person who is intending to learn a job and then to start their own business.

Sometimes artisans employ wage workers who are not apprentices. However, even in this case there is a huge difference between a capitalist company and a handicraft company, as there is no division of labor. The relationship between the "owner" and the "workers" is very close, they use the same equipment, share the same working conditions and the same environment. Workers are often responsible for full production processes, and they have a clear view of most of the structure and operations of the Company.

Artisans usually leverage on their close relationship with customers, on their flexibility in operations, on their cleverness and specialties.

Freelancers and Independent Professionals only rely on their time, skills, and imagination. They try to maximize the return of their time allocation and they leverage on their inner abilities. They may own some means of production, but their main Assets are their own skills, ideas, and capabilities and they also are not capitalistic companies.

There are also a large number of diverse "agents" which act as economic players and belong to the economic system but are not companies. Artists or Associations, for instance, can provide and sell goods and services, but officially they are not "companies".

Non-Capitalistic Companies in a Capitalistic System

Artisans, Artists, Independent Professionals and Freelancers can live and work in a capitalistic system and be a part of it, even if they are economic agents not organized in a capitalistic way, but although they have been on the scene for thousands of years, the capitalist vision was able to set many of the rules on how to organize markets and capitalistic companies could seize the predominant position of artisans in producing objects.

Stott (1996) claims that "collectively, artisan historians portray capitalism as having a strongly adverse effect on craftsmen."

Artisans existed well before capitalism became one of the mainstream economic ideas, as there are archeological proofs that artisans operated in a similar way to modern ones over a thousand years BC, and artisans almost had the monopoly on making goods until the end of the eighteenth century.

Since the beginning of nineteenth century, two elements started changing the economic system, first in the UK and later in the USA: the progressive spread of the mechanization of production processes and the "Transport Revolution", also known as the "Market Revolution".

Both elements required huge investments and were mainly financed by public investments, banks, merchants and owners of huge capitals, most of them involved in slave trade (Harley 2013).

Until the transport revolution occurred, many artisans were trading and already in a market economy, but they were limited in their commercial action, and at the same time protected from competitors, by poor, slow, expensive transportation. George Rogers Taylor (1951) claims that in 1815 in every village there was a local and self-sustainable economic structure aimed to the production of goods. When an extensive network of rails, roads and canals started linking together cities and villages and enabled cheap and fast long-distance hauling, long distance trading became possible.

Taylor (1951) underlines how the division of labor worked at first as a main element for leading the transition to the capitalistic system of industries on which artisans formerly had a monopoly. An example is men's coats production in the mid-19th century. Traditionally, coats were made in two main stages: cutting and sewing. The first stage required high skill, the second step was a time consuming but repetitive job. In rural areas, coats were cut by country Tailors, and then sewed at customer's home by females of family, resulting in a long, expensive process. Clothiers based in cities could employ just one skilled cutter and many low-paid sewers, resulting in a fast and cheap production of ready-made garments. This high-volume production of ready-made garments were addressed to city customers, but then could be sent in a cheap and fast way by train to country villages, and sold in general stores for a fraction of the price of traditionally made coats.

So, at the beginning the capitalistic system gave the opportunity to artisans to become more structured and grow as companies, but the counterpart was that as the production became more and more difficult, equipment became more and more expensive and this also meant that the sum needed in order to start a new enterprise became increasingly large.

As Liza Beth Lubow (1987) underlines in her study of Boston carpenters, while in 1790 45 percent of journeymen became Masters, in 1825 just 11 percent of journeymen became Masters, as capital needs where increasing and repaying debts had become more difficult.

Stott (1996) reminds that "several crafts were largely unaffected by change during the early nineteenth century", like construction trades, and also that mechanization required more skilled workers, but step by step the capitalistic system evolved in a way mass production started to move in giant production sites far from final customers.

In a world dominated by huge capitalistic companies, artisans, freelancers and other independent workers could mainly operate in very local and often small-scale businesses, and maybe interact with the big players as suppliers just for very definite and specific products and services.

This is highly related to the industrial revolution and the way most goods are produced. Other approaches to production have been explored over time, but the mainstream idea became that most of goods and services can only be produced in huge plants, with specialized equipment and an organizational division of labor. Almost all of the huge companies are organized on a pyramidal way, with complex management system, top-down governance, division of labor, separation between owners and users of means of production.

This led to a progressive "proletarization" of highly skilled artisans and to a decline of their living conditions (Stott 1996).

Artisans, Freelancers and other independent workers not only are unable to realize such complex goods and services, but they also have been for a long time, and still are, highly dependent on Capitalist Companies in order to get equipment, raw materials, advanced services, energy, fuel, means of transportation, and so on.

Since the beginning of the XIX Century, huge capitalistic companies ruled the markets in Western Countries, and SMEs and Artisans survived as they were able to provide high specialized goods, to realize small scale productions and to serve local markets.

These players occupied the marginal niches of the market and this ensured their survival for a while, but in the last few years markets changed in a way to affect their specific assets and to weaken their position.

The lifetime of ideas and goods dramatically fell, and this made it impossible to cover costs for research equipments and production lines. Then, globalization gave the opportunity to huge companies to sell directly to their final customers, at incredibly low prices.

This led to a crisis of non-capitalistic local companies like Artisans and SMEs, which are not competitive towards capitalistic globalised companies anymore.

The loss of artisans is something that was perceived as a social and cultural instance too. To lose artisans means to lose the specific skills they have.

"A real craft has the skills attached, if we lose them, we lose the skills attached to them" (Nichols, A., Personal Communication, February 23, 2013).

Without questioning the cultural and social high value of artisans, or the role of capitalistic companies, the interesting point for the purpose of this work is to give an explanation on how it is possible to build an economic non-capitalistic system that can even work together with capitalistic one.

Crowdworkers propose a model of doing business which is not capitalistic, as it's based on different pillars, but that at the same time is not antagonistic. It's just different. Crowdworkers are in between artisans, entrepreneurs and freelancers. They show some characteristic from each of these categories, but their link with the world of "Makers" and their attitude to making things on their own propose them as descendants of artisans. They are entrepreneurs, but they manage money, work, and means of production in a way that is comparable to that of artisans. This is interesting, as artisans are a class of economic agents which always had a dual relationship with Capitalism. While artisans often exploit capitalism as a business opportunity, capitalist companies are historically antagonists of the local artisan systems and the raise of Capitalism usually means a decline of artisanal activities, but artisans have been coexisting with modern capitalism for two centuries, and this is the demonstration that the coexistence of different economic models in the same society is possible.

Crowdmakers are definitely economic agents structured in a non-capitalistic way. Their way to manage material and immaterial assets and means of production based on sharing private properties to gain collective growth puts them in a different position from all the other past economic theories, but this doesn't mean they can't interact and exchange with traditional companies.

From this point of view, the most important point is, therefore, to fully understand the theoretical framework that forms the background to the typical organizational model of crowdworkers and shared work spaces, in particular the meaning that in these contexts is taken by the idea of exchange and sharing.

TOWARDS AN ECONOMY BASED ON THE IDEA OF COMMUNITY

The idea of sharing and exchanging isn't new and is a constant concept always found in economic systems. Every country and every culture developed this idea in different ways, with different degrees of integration on overall social systems.

One of the main economic trends that are growing at international level is certainly the sharing economy. The sharing economy is ideally based on the concept of exchange of goods and services between peers favoured by a platform, but it is at the centre of a heated debate that focuses to a large extent on the role of these platforms. The second part of the chapter fits into this debate, linking the concept of exchange implicit in Co-Manufacturing with the presence of a community.

Therefore, it is important to understand whether sharing activities typical of shared workspaces can or cannot be included in those of the sharing economy, or whether there is a mere assonance in terms, without a concrete conceptual similarity.

Sharing Economy: Criticisms and Different Points of View

The Sharing Economy is often presented as the economy of access to resources, instead of ownership, and is often related to an economy based on on-line platforms that allow a direct link between users and providers of a service.

Hamari, Sjöklint & Ukkonen (2015) define Sharing Economy as "the peer-to-peer-based activity of obtaining, giving or sharing the access to goods and services, coordinated through community-based online services". After mapping 254 collaborative consumption services, they underlined two main modes of exchange: the access over ownership and the transfer of ownership. The access over ownership may take form as renting, as in services like *renttherunway* or *airbnb*, and implies a monetary transaction. Access over ownership may also take form as lending, with no monetary transaction, like in services as *couchsurfing*.

The Transfer of ownership can take form as Swapping, Donating, or Purchasing of used goods. In case of Swapping and Donating, as in services like *Swapstyle* and *Freegive*, no monetary transaction is implied, while in case of purchasing of used goods, like in services as *Thedup*, there is also a monetary transaction.

Yaraghi & Ravi (2017) recall that a *Pew Research Center* report from May 2016 found that 73 percent of Americans were unfamiliar with the term "sharing economy", but 72 percent had used a "shared and on-demand online service", including a wide range of other services as buying second-hand goods on sites such as *eBay* and *Craigslist* and expedited delivery services such as *Amazon Prime*. *Airbnb*, that provides room sharing service, had 11 percent penetration.

Airbnb is a peer-to-peer activity of the type "access over ownership", which takes form as renting a room or an apartment, making a reservation across an on-line platform or an app and giving a percentage for this service to the owners of the platform. A major criticism is that these services put unprofessional providers of services against professional ones, but with fewer constraints and with possible safety issues for end users. Langley & Leyshon (2016) give examples of such criticisms, that see this kind of services as "precarious employment at best, and is certainly not accompanied by the kinds of benefits (e.g. health insurance, pensions) which normally accompany permanent employment."

Another criticism is that Sharing Economy is just a form of "Platform Capitalism" (Lobo, 2014), or "netarchical capitalism" (Kostakis & Bauwens,

2014). In this view, platforms are just the way huge capitalistic companies provide a service, using very small and weak economic agents as subcontractors and keeping in their hands access to network infrastructure. In this view, these economic agents aren't independent workers, but just precarious workers trying to make ends meet until a better job comes along, if any should ever come. This kind of criticism is well expressed by Langley & Leyshonis (2016), when they say that "The degradation of conditions of work that may accompany the rise of the platform is perhaps best illustrated by the rise of the sharing economy platform Uber, which describes itself as a 'transport connection company' that unites 'riders' with (self-employed) 'drivers'. As an intermediary, Uber takes its cut from the fares earned by drivers. Uber's appearance in the taxi markets of cities has mainly been met with hostility from incumbent, licensed taxi companies who are subject to city-based regulations designed to calibrate demand with driver numbers and to protect health and safety."

For many observers, like Friedman (2014), Platform Economy is fostering the rise of the so called "Gig Economy", dominated by short term and casual contracts. This form of job parcelling can be read through what Bernhardt (2014) recalls:

A common intuition about rising inequality in the US is that the reorganization of work and production has contributed to the undermining and erosion of labour standards. The argument is that employers have externalized work as a way to cut wages, reduce benefits, and evade or escape legal responsibility for their employees. This is also a concern about the future– that New Deal laws and policies are inadequate to protect workers in 21st century jobs, whether in the form of temporary work, part-time jobs, subcontracting, or the use of independent contractors.

Cherry (2016) points out that "Some processes of micro labor involve computer based work that is performed wholly in cyberspace, where work is broken down into its smallest constituents elements." while Natour (2017) focuses on the fact that "Gig or on-demand economy, is transitioning from a model based on permanent employees to one of independent contractors or "micro-entrepreneurs" using new platforms to bid for and complete multiple gigs", and this can be related to what Bernhardt (2014) says on nonstandard work, where "nonstandard work arrangements are typically defined by researchers as departing from the standard employment relationship on at least one dimension: (1) the job is temporary, (2) the job is part-time, (3) the worker is employed by an intermediary, or (4) there is no employer at all."

According to Bernhardt, Gig Economy is just a form of nonstandard work conveyed by on-line platforms, but on the other hand there are criticism to this position based on common sense and well expressed by Calloway (2016), when says that "The thing is, if you're working 40 or 50 hours a week with one company, that looks a lot less like a gig and a lot more like full-time employment."

In this view, Sharing Economy seems to be not more than a way for corporations to take advantage of the precarization of otherwise regulated professions, and also the innovative aspect of Platform Economy can be questionable. For example, at first sight an examination of a popular service like Airbnb doesn't seem to reveal any major innovation compared to the past. Access to resources is the base of every rental agreement. Also booking a hotel room is a way to have "access" to a resource, like a place where to sleep for some time in a city, instead of "ownership", as the room is rented and not bought.

Platforms create a direct connection between users of a service and many small providers. But also this business model is nothing new. In many European countries, it's very common to see advertisements outside private houses offering rooms to rent. It's a one-to-one trade and usually in small touristic cities there is one or more agencies that just do what *airbnb* does. These agencies list all the private apartments from the surrounding area that have an available room and put in communication apartment's owners and tourists, taking a percentage of the rental price.

So, one could argue that the Sharing Economy is a way to scale former business models to the advantage of few companies holding the needed infrastructure.

To fully understand the deep implications brought by the Sharing Economy to the way everyday business is organized, what is needed is a better understanding of the real innovation factor brought by some services referring to Sharing Economy.

Langley & Leyshon (2016) notice that "platforms are, at once, techno-cultural constructs and socioeconomic structures" and "analytical attention should be given to the contingent configuration and consequences of the platform as a discrete mode of socio-technical intermediary and capitalist business arrangement." Moreover, according to them, platforms on the one side "mobilize code and data analytics to compose immanent infrastructures" but on the other side they also "enroll users through a participatory economic culture". In their view, platforms "seek to facilitate and capture value from the interactions and circulations of Web 2.0" and the real distinctive aspect is that "platform intermediation targets the ostensible opportunities offered by network effects and the so-called 'co-creation' of value between users". The role of platforms to intermediate network effects means that "platforms are not simply in the business of intermediating connections, but of actively curating connectivity".

Summarizing, the distinctive aspect of the Sharing Economy is not just "what" is transacted, the kind of service that are provided, but "how" this happens, and under this aspect a key point is the quality of connections within a network of people who share the same participatory culture. Starting from this point of view, the authors argue that what really changes with Sharing Economy and services like *airbnb* is the idea of "community", as the community is the environment within which relationships happen and the exchange of resources can take place.

So, a key point to understand how the Sharing Economy works is to understand how communities work and evaluating the strength and the quality of the community of a community-based service is a way to evaluate the solidity of the service itself and to understand if that service is or isn't part of the Sharing Economy.

The Classical View of the Sense of Community by McMillan and Chavis as a Key to Understand the Sharing Economy

Both in Europe and in the United States there is a vivid ongoing debate on the idea of "Community". This includes key aspects like what is a community, what does it mean to be part of it, and how local and global communities change and are affected by global phenomena like migrations, gentrification of cities and the rise of on-line communities.

Although this issue is extremely interesting and of great importance, the authors do not want to contribute here to the debate over the idea of community, nor to stress all the aspects of the debate over peculiarities and differences between on-line communities and physical communities, but will rather focus on what is a community in relationship to the Sharing Economy and on the implications and importance of a community when talking about Sharing Economy.

To the extent that is enough for the purposes of this book, the authors will refer to the classical theories by Mc Millan & Chavis and will assume the common idea of a community as a network of people who all share same common interests, or a common cause, and who feel to be part of it.

According to McMillan (1976), the Sense of Community is "a feeling that members have of belonging, a feeling that members matter to one another and to the group, and a shared faith that members' needs will be met through their commitment to be together."

Later Mc Millan & Chavis (1986) identify four elements which compose the Sense of Community: Membership, Influence, Integration and fulfilment of needs, Shared emotional connection.

As they assume, "Membership is a feeling that one has invested part of oneself to become a member and therefore has a right to belong (Aronson & Mills, 1959; Buss & Portnoy, 1967). It is a feeling of belonging, of being a part (Backman & Secord, 1959). Membership has boundaries; this means that there are people who belong and people who do not. The boundaries provide members with the emotional safety necessary for needs and feelings to be exposed and for intimacy to develop (Bean, 1971; Ehrlich & Graeven, 1971; Wood, 1971)." while "Influence is a bidirectional concept. In one direction, there is the notion that for a member to be attracted to a group, he or she must have some influence over what the group does (Peterson & Martens, 1972; Solomon, 1960; Zander & Cohen, 1955). On the other hand, cohesiveness is contingent on a group's ability to influence its members (Kelley & Volkart, 1952; Kelley & Woodruff, 1956)."

Mc Millan & Chavis summarize the concept of influence in these four points:

1. *Members are more attracted to a community in which they feel that they are influential.*
2. *There is a significant positive relationship between cohesiveness and a community's influence on its members to conform. Thus, both conformity and community influence on members indicate the strength f the bond.*
3. *The pressure for conformity and uniformity comes from the needs of the individual and the community for consensual validation. Thus, conformity serves as a force for closeness as well as an indicator of cohesiveness.*
4. *Influence of a member on the community and influence of the community on a member operate concurrently, and one might expect to see the force of both operating simultaneously in a tightly knit community.*

The third component of Mc Millan & Chavis definition of sense of community is Integration and Fulfilment of Needs, which, they say, "translated into more ordinary terms, is reinforcement. Reinforcement as a motivator of behavior is a cornerstone in behavioral research, and it is obvious that for any group to maintain a positive sense of togetherness, the individual group association must be rewarding for its members."

This is extremely important in understanding the link between the idea of community and Sharing Economy, as Mc Millan & Chavis stress out a key aspect:

The main point is that people do what serves their needs. But this leaves questions unanswered: How do people prioritize their needs, especially after meeting the basic survival needs? What creates a need beyond that of basic survival? Reinforcement as an organizing principle seems blind and directionless unless it is complemented by other concepts. One such directing concept is shared values. Our culture and our families teach each of us a set of personal values, which indicate our emotional and intellectual needs and the order in which we attend to them. When people who share values come together, they find that they have similar needs, priorities, and goals, thus fostering the belief that in joining together they might be better able to satisfy these needs and obtain the reinforcement they seek. Shared values, then, provide the integrative force for cohesive communities (Cohen, 1976; Doolittle & MacDonald, 1978). Groups with a sense of community work to find a way to fit people together so that people meet the needs of others while meeting their own needs. (cf. Riley, 1970; Zander, Natsoulas, & Thomas, 1960).

The following summarizes the role of integration and fulfilment of needs in a sense of community: 1. Reinforcement and need fulfilment is a primary function of a strong community. 2. Some of the rewards that are effective reinforcers of communities are status of membership, success of the community, and competence or capabilities of other members. 3. There are many other documented needs that communities fill, but individual values are the source of these needs. The extent to which individual values are shared among community members will determine the ability of a community to organize and prioritize its need-fulfilment activities. 4. A strong community is able to fit people together so that people meet others' needs while they meet their own.

According to Mc Millan & Chavis, the fourth key aspect to have the Sense of a Community, is to have a Shared Emotional Connection within members. They notice that

A Shared Emotional Connection is based in part on a shared history. It is not necessary that group members have participated in the history in order to share it, but they must identify with it. The interactions of members in shared events and the specific attributes of the events may facilitate or inhibit the strength of the community.

The following features are important to the principle of shared emotional connection:

1. ***Contact Hypothesis:*** *The more people interact, the more likely they are to become close (Allan & Allan, 1971; Festinger, 1950; Sherif, White, & Harvey, 1955; Wilson & Miller, 1961).*
2. ***Quality of Interaction:*** *The more positive the experience and the relationships, the greater the bond. Success facilitates cohesion (Cook, 1970).*
3. ***Closure to Events:*** *If the interaction is ambiguous and the community's tasks are left unresolved, group cohesiveness will be inhibited (Hamblin, 1958; Mann & Mann, 1959).*
4. ***Shared Valent Event Hypothesis:*** *The more important the shared event is to those involved, the greater the community bond. For example, there appears to be a tremendous bonding among people who experience a crisis together (Myers, 1962; Wilson & Miller, 1961; Wright, 1943).*
5. ***Investment:*** *This feature contributes more than just boundary maintenance and cognitive dissonance. Investment determines the importance to the member of the community's history and current status. For example, homeowners who have invested money and time in their part of a neighbourhood are more likely to feel the impact of the life events of that community. Similarly, persons who donate more time and energy to an association will be more emotionally involved. Intimacy is another form of investment. The amount of interpersonal emotional risk one takes with the other members and the extent to which one opens oneself to emotional pain from the community life will affect one's general sense of community (Aronson & Mills, 1959; Peterson & Martens, 1972).*

6. *Effect of Honour and Humiliation on Community Members: Reward or humiliation in the presence of community has a significant impact on attractiveness (or adverse- ness) of the community to the person (Festinger, 1953; James & Lott, 1964). Spiritual bond: This is present to some degree in all communities. Often the spiritual connection of the community experience is the primary purpose of religious and quasi-religious communities and cults. It is very difficult to describe this important element.*

Mc Millan & Chavis continue recalling Bernard (1973), who calls this factor "community of spirit," and link this factor to the nineteenth-century concept of volkgeist (folk spirit). Mc Millan & Chavis state that "The concept of soul as it relates to blacks and its role in the formation of a national black community is an excellent example of the ole of a spiritual bond. They [blacks] had a spiritual bond that they understand and that white people could not. Soul was an indefinable, desirable something; black people had it but white people could hardly aspire to it. It was the animating spirit behind their music, their dance, and their styles. It even expressed itself in their taste in food, their language, and their speech. Not even all black people shared it. Those who rejected their blackness did not. Bernard, 1973, p. 130) This element of shared emotional connection can be traced through Tonnies' (1957) use of the term gemeinschaft: a social unity based on locale. According to Konig (1968), gemeinschaft's root, gemeinde (local community), had a long-time original application as "the totality of those who own something in common" (p. 15). Cohen (1976) found this in the related concept of the Bund. Neither gemeinschaft nor Bund nor shared emotional connection as presented here includes the requirement of a small-scale local community. Kasarda and Janowitz (1974) demonstrated that "increased population size and density do not significantly weaken local community sentiments" (p. 338), which further aids us in understanding communities that are not bounded by location."

Starting from the assumptions of Mc Millan & Chavis, it can be stressed that this is the huge difference between community-based services related to the Sharing Economy and other kind of services.

Shifting from theory to a practical example, *Airbnb* is different from a booking service like *booking.com*, and the difference is the community. On *booking.com* there is a platform that puts together needs and offer. There are professional providers of services, like hotels, and customers, who both don't feel to be part of the same group.

In this sense, the community of *airbnb* is made of people who both offer their rooms to others and use the service as users too.

Providers and users are not on two sides of a barricade. They are not counterparts. They are exactly the same people, who share the same values and belong to the same community.

A community made out of people who believe in travelling as a way to meet locals and who want to feel the taste of living as a local. Within the community, the shared values are meeting new people, giving and receiving advices and hints, feeling at home also if abroad.

If seen under this point of view, a service like *airbnb* is not in competition against hotels and formal tourist accommodations. The idea is to travel and being hosted by a network of people as if they were friends, and the money exchange is more a small financial contribution for having troubled the host rather than a proper payment for a service. People who list their rooms often have a job that has nothing to do with tourism and what they gain from *airbnb* is just a small extra amount of money.

When the platform works this way, then the Sharing Economy reveals its innovative aspects, and it's easy to see that this happens when the community is activated on the four elements which compose the Sense of Community identified by Mc Millan & Chavis (1986), Membership, Influence, Integration and fulfilment of needs, Shared emotional connection.

Of course, in *airbnb* there are many cases that go far beyond this rule. In some cases, rooms are listed by people who really need this money, and in that case it's Gig Economy. In many touristic cities, in Italy for instance, there are people who own several apartments and rent them as their primary job. These people are not part of a "community", they are just renting rooms as a professional service. There is no sharing of emotions or values between who rents and who offers in this case, it's just capitalism.

So, in a way it can be said that the Sharing Economy is a behavioral economic model, as it's closely related to the intentions and personal feelings of the people who play an active role in it, and whether these people behave in a way that reinforces the structure of the community, or just use the network to gain personal advantages following a different economic model.

The sharing Economy requires something to share within a community and the assumption is the existence of the community. So, it's really important how this community is shaped and ruled, and on-line platforms easily show some weak points under this aspect.

For instance, looking again at the *airbnb* case, there is no advantage in being both a user and a provider of the service. There is a rating system that gives an idea of the coherence between what a user claims and the reality, but it works fine also if the user is a touristic accommodation and not a private person.

There is no advantage in balancing the roles of "user" and "provider".

This means that a user who hosts people for thousands of nights without ever being hosted has no disadvantages coming from this behavior.

To some extent this is inevitable, due to the particular nature of services offered on *airbnb*. It's clear that not everybody can host people, and it's also clear that nobody can be forced to travel, but on the other side this can lead to attitudes and behaviors that go against the basic principles of sharing resources within a community and this could be fatal for *airbnb*.

For instance, there can be people who own and rent entire buildings, with a behavior more similar to an unofficial Hotel than to a member of a community. Brescia (2016) proposes an articulated way to regulate Peer-to-Peer Economy, in order not to hinder with innovation and at the same time to protect customers and market competition, and many cities around the world are defining regulations to prevent such situations.

This aspect should be further clarified looking at other two well known cases of business based on the Sharing Economy: *Bla Bla Car* and *Uber*.

Bla Bla Car is a community of car drivers who offer rides in their cars when travelling to their own destinations. In this case, the aspect of "community" is quite clear. The platform is used only to put in contact members and routes and prices are self-organized by members of the community. Prices are very low, just enough money to share travel expenses and the contact between who offers a ride and who takes it is very close. *Uber*, on the other hand, has a highly centralized booking service, and there is a clear distinction between riders and customers.

In that regard comes the criticism of Asher-Schapiro: "drivers aren't *partners* — they are laborers exploited by their company. They have no say in business decisions and can be fired at any time. Instead of paying its employees a wage, *Uber* just pockets a portion of their earnings. Drivers take all the risks and front all the costs — the car, the gas, the insurance — yet it is executives and investors who get rich.

Uber is part of a new wave of corporations that make up what's called the "sharing economy." The premise is seductive in its simplicity: people have skills, and customers want services. Silicon Valley plays matchmaker, churning out apps that pair workers with work. Now, anyone can rent out an apartment with *Airbnb*, become a cabbie through *Uber*, or clean houses using *Homejoy*.

But under the guise of innovation and progress, companies are stripping away worker protections, pushing down wages, and flouting government regulations. At its core, the sharing economy is a scheme to shift risk from companies to workers, discourage labour organizing, and ensure that capitalists can reap huge profits with low fixed costs.

There's nothing innovative or new about this business model. *Uber* is just capitalism, in its most naked form".

While many observers and scholars share this point of view, many others here can argue that the pressure on a steady market coming from new economic actors can lead to positive results. Koopman, Mitchell, and Thierer (2015), state about the *Uber* case that this positive result can be seen mainly under four aspects:

"A. Expanded Range of Goods and Services ", as "The ease of entry and innovation in the online world mean new entrants can provide better options and address problems previously thought to be unsolvable in the absence of regulation. Polls have revealed consumers currently take advantage of sharing economy services primarily because they offer greater convenience, better prices, and higher quality. This is also attested by comparisons of Yelp ratings in almost any major city where ride-sharing firms operate.

B. Expanded Information", as "The best that could be hoped for in the pre-Internet era was that consumer watchdogs, competition between firms, and brand "goodwill" would be enough to safeguard consumer welfare. But, the Internet largely solves this problem by providing consumers with robust search and monitoring tools to find more and better choices. These tools lower both search costs and transaction costs associated with commercial interactions. Moreover, unlike regulatory solutions, these market-developed tools cannot be captured. Online e-commerce and the sharing economy developed thanks to these new realities. C. Consumer Empowerment via Reputational Feedback Mechanisms", as "information technology has facilitated the creation of countless reputational feedback mechanisms across the online ecosystem—such as product rating and review systems—that give consumers a more powerful voice in economic transactions. Before the Internet, "reputations travel[ed] haphazardly [through] word of mouth, rumor, or the mass media," but first-generation e-commerce sites like eBay and Amazon helped blaze the way for far more robust reputational feedback systems." They also point out that "Today, almost all sharing economy firms depend on these reputational feedback mechanisms to establish trust between suppliers and consumers. David D. Friedman notes "reputational enforcement works by spreading true information about bad behaviour. People who receive that information modify their actions accordingly, which imposes costs on those who have behaved badly." Sharing economy entrepreneurs have developed a number of other monitoring mechanisms to ensure quality". And, "D. Self-Regulating and Other-Regulating Markets the combination of these factors results in a powerful check on market power or abusive behaviour. The reputational incentives at work require firms to constantly seek ways to satisfy rapidly evolving consumer demands and to gain (and keep) consumers' trust." "And, because sharing platforms have opened traditionally cartelized industries to new competition, they have also permitted firms to regulate one another's behaviour. Competitive firms are often quicker than regulators to point out the substandard service of their rivals. The result is reasonably well-functioning, self-regulating markets with strong checks on improper behaviour".

From Sharing Economy to Community Economy

Even if it is not possible to identify a unitary line by scholars in judging the Sharing Economy as a whole, what clearly emerges is that there is a growing separation between those experiences that focus on technology to provide cheaper and faster services based on apps and web platforms, and those experiences that increasingly focus on the idea of community and the sharing of values within it.

A number of experiences started over the years which weren't based on an app or an on-line service, like the *Leila Project*, a "library of things" started in Berlin in 2011 that allows members of a community to borrow household items from each other.

The *Leila Project* refers to the idea of collective ownership, but the idea is developed in a such a way that no internet or membership fee are required.

Pumpipumpe is a project with the goal to create local Sharing Communities, started in Switzerland and aiming "to make sharing-friendly neighbours and their objects visible and so to promote the sharing of consumer goods. With *Pumpipumpe* this does not happen on the Internet, but with little stickers on the letterbox, where neighbours and local residents walk past every day. That way you are able to directly contact each other to borrow a bike-pump, drill or a fondue set, get to know each other better and have to buy even less equipment."

Another interesting case is offered by the so called "*Free Shops*", that are on the increase in almost every town of Central European Countries, like Germany. These shops are places where people can bring their used goods, and everything can be taken for free. These shops are based on the idea to offer a chance to people from a small local community to have a place where they can bring their used goods and take something left by someone else. These places are "physical platforms", as they are places where members of the community meet and perform the exchanges. There is no difference between who brings used goods and who takes something, as they are exactly the same people. Used goods are seen as "common goods" that are made available to the local community and everybody can benefit from these common goods. Living expenses and running costs of the free shops are usually covered by the community as there is an "honesty box" where everybody can put some money, as a free offer.

Also in this case, the inner coherence of the community is the main issue.

The system works only if there are many people who bring their used goods and take others' used goods, as this grants a high rotation of items and an overall good quality. If people who bring and people who take are the same, also the style and the kind of available goods will be homogeneous and interesting for the members of the community. If people who bring and people who take are part of different groups, *Free Shops* become something different and less interesting. For instance, if there are a lot of homeless and poor people who take goods, *free shops* become something more similar to voluntary charitable where "rich" people bring used stuff and "poor" people take them. Of course, there will always be a fraction of people who just bring used clothes, but who would never wear used ones. And there will always be a fraction of people who just go to the *Free Shop* to take something, like the homeless or very poor people.

This is not a problem, if these behaviors remain marginal and most of the community is made up of people who both bring something and also take something. If there are a lot of people who "take", there will also be many more offers to run the space and this can lead to longer opening times, possibly creating the need of a waged worker to make this possible.

So, the main shield is the strength of the community. If the community is weak, the economic system is also weak. The interesting point is that members of these communities, even when they are involved, are more focused on their personal gains than preserving the community.

Hamari, Juho and Sjöklint (2015) found that in an economic system based on sharing economy "even if the particular motivations of individual participants vary from mainly altruistic to strongly gain-seeking, the sharing economy as a whole remains functional, provided that the benefits for each participant outweigh possible costs incurred through the imbalance of contributions. And, of course, "economic gains" as defined in this study also translate into saving money, which is an understandable motivator for many consumers such as those affected by the recent financial crisis". Moreover, "Collective Consumption has been regarded as a mode of consumption that engages especially environmentally and ecologically conscious consumers. Our results also support the notion that viewing Collective Consumption as a sustainable activity can lead to an increase in participation, but only if by taking this view we increase positive attitudes towards Collective Consumption. Our results, however, also suggest that these aspirations might not translate strongly into action. Expectations as to the diffusion of Collective Consumption might thus be deflated; it may actually be people seeking economic benefits who in the end opportunistically adopt Collective Consumption as one of the modes of consumption. In a worst-case scenario, some users in a sharing economy might be altruistic and share their goods whereas other users may be mostly enjoying benefits from others' sharing. This situation might affect the sustainability of Collective Consumption services in general."

Rules and Governance as a Key to Preserve the Community: The Role of the Platform

Durkheim argues that a certain degree of deviance is a normal phenomenon for any social structure, which becomes negative only when it is accompanied by a state of disorganization in which the whole system of conduct rules loses value. According to this, the most important aspect in preserving the health of the community is how the community is shaped.

So, if it's reasonable to expect some deviant behaviors from the average behaviors within the community, organization is the bank that prevents these deviant behaviors from flooding the community.

The Sharing Economy is based on the idea of "sharing" as a way to make some resources available within a group of people.

In order to make this happen, there are some elements that have to interact. First a group is needed. People who know to be part of a group, who own some resources and want to share them within the group.

Then there is the need of a way to distribute these resources and to share within the group. This implies that some rules are needed to oversee the way these resources are both given and taken, and there is the need to identify an accumulation point for these resources, so that those wanting to give resources know where to put, or to list, said resources and those needing the resources know where to pick, or to retrieve, them.

So, an important element is the presence of a "platform", a virtual or physical place where these resources can be collected and distributed.

The presence of a "platform" is important, for two reasons. The first one is that those who own some resources can put them into the "platform" and those who need these resources can visit the "platform" to find what is needed. The second one is that who runs the platform also sets the norms to rule the exchange of the resources.

Looking at the traditional examples of business services that are based on the Sharing Economy, this two aspects are both clear.

Most platforms are places where who owns a resource can "put" it, and who needs it can "take" it directly from the platform. The contract, the payment, the agreements are performed inside the platform, and the rules are decided by the platform.

So, for sharing economy the first pillar is the availability of the platforms, and the second one is the presence of a community.

As stated before, community is a crucial point in developing an economic system built around the idea of sharing. Sharing requires a deep level of trust and requires a kind of stability regarding the group, that allows for Mutual help, Mutual transfer of skills, and increases the strength to have access, as a group to high level resources.

According to Focardi & Salati (2015),

Within a Multifactory, there are two main reasons for which that put in evidence the need to exchange knowledge:

1. *The solopreneurs, entrepreneurs, SMEs, and artisans who are members of a Multifactory in most cases are startups and come from previous jobs as employees, and/or completely different sectors. This means that they have ideas and most of the technical skills needed to start their own business, but they miss many other skills. So, they face problems they don't know how to solve, and an exchange of knowledge can directly solve them, or help them to retrieve the required information to autonomously find a solution.*
2. *There used to be a time when it was very important to protect an idea, let it grow internally in the company and then to develop an exclusive product/service that could last for years. Nowadays, the life cycle of ideas, products, and services is so fast that it's less important to protect*

them. It's much more important to develop them as fast as possible, in a very reliable and flexible way, to target the market as fast as possible. A single small or medium company, an artisan or a pro should have many good ideas throughout the years, and they can be very different from others, so they can't rely only on their knowledge, skills and experience, but they need an easy, fast, reliable ways to get competencies. If they're part of a system where they're used to an everyday free exchange of knowledge between well-known subjects, they can find the resources they need, saving financial expenses and avoiding the cost of retrieving resources (for instance, to obtain a loan), and the cost of creating a completely new coordinating structure for each single project.

This wouldn't be enough to secure the community, as a community is not just made out of common needs, but also common answers to these needs.

Focardi & Salati (2015) point out that "A crucial issue in a system where free exchange of knowledge is an important aspect of how relationships are regulated between nodes is that whoever spreads competencies, or suggestions, should have some form of non-monetary return.

In most cases, the "knowledge exchange" also requires some labor, sometimes the use of their equipment and machines, and at least the use of an (many times small) amount of time.

Many people, in their private lives, share knowledge for philanthropic reasons, or because it's something that belongs to their values, or beliefs, or just use it as a way to spend some spare time. In this sense, the return of sharing could also be just the proud of being part of a movement, or the consciousness of being useful for someone else. This obviously should not apply to entrepreneurs, professionals, or artisans; at least not as a strategy of growth.

At first sight, the return should be intended as a mere exchange of professional services, a way to offer and get services without money, or a kind of barter: "I have a need, I looked for a skill, I found it, I offer something in change."

This is something that happens, but there is something much more interesting well hidden, and that is the capability of the system to increase through knowledge, sharing the growth opportunities of the Companies that are part of the system.

What happens is that availability of competencies enables and leads people to think in a different way. They know that when they face some problems, or professional challenges, they can count on a wide range of capabilities and skills that are available and reliable, and this constant, continuous availability of resources enables entrepreneurs to think with a wider approach.

It's a buffet of skills. Entrepreneurs know which resources are available and how to get them. Maybe they don't need them for a long time, but they are aware of the availability and this leads them to take on big projects because they know that they will have access to all they will need.

So, when they put knowledge into the system, what they're doing is raising the global potential of the system itself, and if everyone in the system offers something different, the result is that every node in the system can grow faster and steadier.

It's a way to build a system where all the players give their contribution to increase the capabilities of every single node of the system itself.

This is an emerging property of the system, which comes from the interaction between elements of the system, and that increases the possibilities of each node of the system to afford new and bigger projects, which means an increased possibility for success on the market."

According to these remarks, the stability that can be gained in shared workspaces is not linked to the personal value of members but is due to the inner structure of the system.

The sharing structure is the key to imagine a New Way to be an Entrepreneur and links the success of each member to the skills and means of production brought by other members. The structure prevents egoistic behaviors, like taking and non-giving, because if a member doesn't get enough in exchange of what he brings to the others, he will exit the multifactory, according to the behavior expected following observations by Hamari, Juho and Sjöklint (2015).

Given this, it's obviously highly important to understand how tradings and exchanges are regulated, as norms and rules are a key point to preserve the fairness of the environment and to avoid the risk of egoistical behaviors.

This trust mechanism is provided by the presence of an organized structure, the multifactory as well as other kind of shared workspaces, that works as a platform to foster exchanges.

As a platform, a multifactory sets the basic rules for the members to interact, but is just a layer, while the dynamics between members are self-regulated.

On top of this platform there are exchanges and sharing of valuable skills and other assets, but well before any exchange there is a community, because skills and assets are strictly linked to specific members of the community.

This system avoids any risk of becoming a kind of gig economy, as mutual help is given by people to people, whereas can take form as small hints, and sharing is less focused on exchanging services, whereas is focused on a steady and all-round empowerment of all members through the constant presence of other members.

In this system, the focus is less on sharing specific skills and more on caring about others, and therefore authors would prefer to shift the general mainframe from Sharing Economy to Community Economy, as the community is the important aspect.

Actions are in a way secondary, as people don't have to act in any special way. The system is designed in a way that when members focus on their own projects, and grow their own companies, all the system has a beneficial outcome.

ABOUT THE NATURE OF EXCHANGES WITHIN COMMUNITIES OF WORKERS: THE IDEA OF NON-ECONOMIC SERVICES OF GENERAL INTEREST

The idea of community that is found in shared workspaces is strongly linked to the presence of different actors that express similar needs, behaviors and values. This already creates the idea of being part of a group, but then this idea becomes explicit in everyday life through a series of exchanges among the economic subjects that are part of the shared workspace. This series of exchanges take place on a daily basis and involves advices, suggestions, ideas and other elements which constitute a true "set of services" that companies give to each other.

These services are exchanged between members and this exchange of services may seem in some way to be a form of barter, but the idea at the base of Co-Manufacturing and Community Economy is far different from Barter Trading and Swap Economy in general.

As an economic service does not change its nature and remains such even if it is delivered free of charge, the point is not to determine whether or not a payment is due, but if this exchange between companies can represent something completely innovative and of a different nature from any service rendered, for example, by a coach or consultant.

It's extremely important, therefore, to determine whether this generic "mutual aid" between companies is something additional, innovative, not anticipated by the current economic system, or is a simple substitution, with free and amateur advice, of opinions and skills otherwise provided by paid professionals.

The first idea that comes to mind when talking about exchange of services or goods is Barter.

Barter trading is a traditional way to exchange goods and services and in many Countries is well regulated by commercial law.

In Italy, for instance, the barter transaction qualifies as a swap contract (art. 1552 Civil Code), and is ruled by art. 11 and Article 13, paragraph 2, lett. d), of Presidential Decree 633/72. Law states that a company can freely barter its goods or services with other companies, as well as with private persons, and giving goods or services in exchange for other goods or services is considered an alternative mean of payment. Also the exchange of goods against services or services against services is considered a barter.

Under the fiscal point of view a barter isn't different from other transactions involving means of payment that require the use of money, and constitutes a taxable transaction for VAT purposes (Article 1 of Presidential Decree 633/72).

Barter trading between companies is regulated in a way similar to other transactions and the company that gives the goods or services has to issue a regular invoice to the receiving company. A special statement on the invoice has to identify the lack of financial flow, and the transaction is subject to VAT as specified in Articles 2 and 3 of Presidential Decree 633/72.

Each barter must be considered as the combination of two different transactions and every transaction must be assessed separately, and an invoice should be issued for every individual transaction.

Although Barter Trading is widely used within companies, in Shared Workspaces based on collaborative economy the kind of support between companies is something new and different from an exchange of services, as it is a kind of support that comes in a way that would not be possible in the traditional labor or consultancy market.

Collaborative Economy creates a type of exchange that is of a different nature from the traditional customer / supplier relationship.

Focardi & Salati (2015) noted that in a multifactory "There are "old style" artisans, who have huge amounts of experience and know very well how to use specific equipments and machinery, but who don't know anything about marketing or have never used Social Networks.

There are experienced professionals, who have all the skills needed to shape and project an object, but who don't know how to realize a physical prototype by hand.

The "knowledge" all these subjects may exchange can be of several types and can cover a wide range of activities. Just to exemplify, knowledge of new materials and techniques (like 3D modelling and printing, laser cutting, fast prototyping in general), knowledge of traditional techniques (like soldering, woodworking, sewing, and so on), immaterial knowledge (like copywriting, project writing, proofreading, etc), digital knowledge (like web design, SEO, programming) or other technical skills (like video-making, music composing, event organization).

All these people are much more interested in a learning process that allows them to experiment and to try on their own rather that following some set rules. They have chosen a place without any expensive learning programs also because they have [inside] this "maker attitude" that brings them to find, through Internet and through a process of trial and error, solutions to their everyday challenges."

These new entrepreneurs have a mutual relationship similar to the one apprentices used to have towards their master craftsmen. They all own a wide range of skills and knowledge but lack many others. Sometimes those in need do not recognize their need, or still do not know how to formulate it as an explicit request, and those who respond to the need do not know that they are responding to that need. This is similar to what used to happen to apprentices, as they entered the workshop knowing they would need to increase technical skills, but then they realized they needed not only to learn technical skills, but also how to handle suppliers, customers, relationships with local stakeholders and so on.

It is therefore crucial to outline a theoretical and regulatory framework within which to define the nature and mode of such exchanges. It's important to define whether these exchanges are something new, that leans on top of other economic exchanges and doesn't affect other types of economic relationships, like specialized consultancy.

Local Public Services of General Economic Interest and Non-Economic Services of General Interest in Local Area

There is a huge debate in Europe on the definition and boundaries of the services of general economic interest in local area (or local public services of general economic interest) and non-economic services of general interest in local area, as this has direct implications on the way public bodies can interact with providers of such services.

The national Italian legislature appears to have *prima facie* configured, through the most recent legislative measures, local public services of economic interest using the European model, which then needs to be used to determine the perimeter of services of economic interest for the Italian order.

Midiri (2017) recalls that "European law and jurisprudence started by identifying services of general economic interest as those that guarantee wages and revenues and, therefore, generate profits and be placed in the competitive market as comes out of the Commission Communication of 19 January 2001, Services of General Interest in Europe [COM (2000) 580]. In Official Journal C 17 of 19.01.2001, paragraph 30, there is a direct reference to activities which "do not make profits and do not to carry out an industrial or commercial activity " as this comes from sentences from the Court of Justice in Case C-109/92 Wirth c. Landeshauptstadt Hannover, and in Case C-263/86 Belgian State c. Humbel and Edel in 1988 p. I-5365, which states that "It should be noted here that, pursuant to art. The first paragraph of Article 60 of the EEC Treaty covers services in the chapter relating to services solely as 'services normally provided for remuneration'. The notion of remuneration is not explicitly defined in Art. 59 et seq. Of the EEC Treaty, however, its legal scope can be inferred from Art. The second paragraph of Article 60 of the Treaty provides that the services include, in particular, industrial and commercial activities and crafts and leisure activities. The essential feature of remuneration must therefore be found in the fact that it constitutes the economic remuneration of the consideration in question, which is generally agreed between the lender and the recipient of the service." Justice Court, 17 February 1993, Joined Cases C-159/91 and C-160/91 Poucet and Pistre c. AGF and Cancava, in 1993, p. I-637; Communication Ult. cit., sub Annex II, Definitions."

For what concerns the non-economic services, Midiri also reminds that In COM 1996, the scope of non-economic services was limited in a narrow sense by identifying them in activities that concerned not only the prerogatives inherent the exercise of public powers (such as security, justice, etc.), but also those intrinsically non-economic, such as the school system for what concerns compulsory education, or social security systems.

This criterion was applied in many national law systems and also in Italy at the beginning this idea of service of general economic interest coming from European legislation was just taken and transposed into the National legal order, linking to the above mentioned European approach the distinction between services of "economic importance" and "services of no economic relevance" in the Articles. 113 and 113 bis of Legislative Decree no. 267/2000. What

Gallo says referring to the Case C-475/99 Ambulanz Glockner [2001] ECR I-8089 discussed in front of the European Justice Court, when law identified an activity as of economic interest under an objective "comparative" criterion, which means that an activity is of economic interest when it is placed in a market in which competition between operators can be altered, and this is valid also if this market is just a potential one.

This criterion, which has been often cited in European and national Courts, is very important as it doesn't refer to the legal substance of an entity to define whether it's performing an economic activity or not, but to its behavior in that specific case.

The European figure of service of general economic interest was then enriched with further specifications. With the 2006 Communication, it was stated that "with regard to competition law, the Court has ruled that economic activity must be considered as any activity consisting of offering goods or services in a given market by an undertaking, irrespective of the legal status of that entity and its financing arrangements ". The consequence is that "if the public authorities decide to entrust the mission to an external partner or to cooperate with the private sector with a partnership, the Community rules on public procurement and concessions can be brought into play" and each case must respect the principles of transparency, equal treatment and proportionality. If an authority carries out a "public financial compensation" to cover the costs of carrying out the mission and which would not have been incurred by an entity operating solely on the basis of market criteria, the Commission may prohibit rules prohibiting State aid.

In this sense, the status of the body that provides a service is irrelevant in order to qualify the service as relevant or not. This criterion was for instance followed by the State Council (2013), which stated, for voluntary associations, that the absence of a profit does not exclude the presence of a competitive economic activity: "Voluntary associations may be admitted to public tenders as social enterprises, to which the d. lg. March 24, 2006, no. 155, has recognized the legitimacy of exercising a stable and principal economic activity organized for the production and exchange of goods or services of social utility aimed at achieving general, though not lucrative, purposes of public interest; as a result, the absence of a lucrative purpose does not preclude such persons from exercising economic activity and being considered economic operators by satisfying the necessary requirements to be qualified as entrepreneurs, suppliers or service providers, and thus consistently with Article. 1, par. 8, Directive No. 2004/18 / EC, for which the terms of the contractor, supplier and service provider designate a natural or legal person or a public body or a grouping of such persons and / or bodies offering on the market, respectively, the execution of works and / or works, products or services."

Midiri also remembers the Opinion of the Council of State, Opinion No 01075/2016, p. 12, "More recently, the Court of Justice has held, after having considered as irrelevant for the purpose of providing the services on the market the fact that the operator is a non-profit-making body (and that its staff performs voluntary service), that the economic nature of a service can not be ruled out and its submission to the Community framework of public contracts (referred to in Directive 2004/18) only because the remuneration is limited to reimbursement of the costs incurred20. In this case, the perimeter of the notion is determined by a joint application of the Treaty, and its art. 49, 56, 105, 106 and the most recent procurement directive." Midiri, coming back to the Italian notion of service of economic interest, also says that "the Unic Text, art. 1 and 2, implies in the national system of local public services the European concept of services of general economic importance, overcoming the definition of 'economic relevance'", as, for instance, the Constitutional Court in its judgment no. 325 of 2010 distinguished the two figures of services, also accepting the objective benchmark of the potential market, according to which there is a competitive market even if this is not fully developed, but there are objectively the conditions for developing it.

A Boundary Between Public Services of General Economic Interest and Public Services of Different Nature

So, national Italian laws assume the comparative criterion, according to which, it is sufficient to suppose, on the objective level, that there is a competitive, even potential, market, not finding a discretionary appreciation of the administration.

This way to define a boundary between public services of general economic interest and public services of a different nature makes this line of demarcation between public services of economic interest and private services more flexible, as they both fit into the field of the competitive market.

This idea of services of general economic interest is very market oriented, and this makes it very interesting to deal with, as it defines clear and strict boundaries to what public bodies can or cannot do in order not to break the European Laws to protect free competition on the market.

This is very interesting when it comes to local public bodies, like municipalities, and how they can deal with economic services that are run at a local level when it comes to giving them some kind of support.

For instance, a local small municipality that wants to support a local bed and breakfast could be performing an illegal action, even if on its territory there are no local competitors, as there could be the theoretical option that a private investor could want to build a hotel in that region. The same could be argued with a municipality that wants to support an emerging economic environment as a Maker Space or a Coworking.

Starting from the idea of services of general economic interest, following Midiri (2017) it is possible to better understand the idea of services of general economic interest in local area. To this end, it is also necessary to reflect on how the legislator has outlined the figure of services of general economic interest in the local area. At first examination it seems to be possible to find that the legislator intended to translate the European view of services of general economic interest as it is coming from EU laws and it is a specification of services of general interest. But in importing the concept into Italian order, as emerges from the texts mentioned, it seems to have added some additional features which may have enriched the idea and expanded the ' the scope of application. To the notion adopted, the Unique Text follows a set of principles and rules applicable to local public services of general economic interest and not to public services of general economic interest.

In Italy it's possible to refer to the definition of Services of general economic interest in local area that can be retrieved the legislative decree n. 175 of 19 August 2016, and Midiri states that the notion of services of general economic interest in the local area can only come from both letters h) and i) of art. 2 of Legislative Decree 19 August 2016, no. 175. By combining the two letters, it can be said that such "services of general interest disbursed or likely to be disbursed on an economic basis on a market which the metropolitan municipalities or cities, within their respective competencies, take as necessary to ensure the satisfaction of the needs of the reference collectives so as to ensure the homogeneity of development and social cohesion ".

Midiri (2017) again says that "the definition of the two notions is therefore particularly relevant because they would not be applicable, inter alia (excluding all organizational and functional provisions), or the recruitment mechanisms of the service by the local authority (direct trust in particular circumstances or award to the market by tender) 5, nor the safeguard mechanisms in the broad sense, of course more than court jurisdictions (service contract and service cards), as the fifth Title of the Unique Text, which goes from art. 21 to art. 32, sets out a series of provisions to ensure transparency and consumer protection. Consider, for example, rules on a service contract, with its mandatory content and service cards. There are also rules requiring checks to allow the entity entrusted, inter alia, to verify compliance with the service contract." And "As per art. 4 of the Unique Text the purposes and the principles of management can be counted among the aims of "affirming the centrality of the citizen in the organization and production of local public services of general economic interest, also by promoting forms of active participation ... To this end, respecting the Constitution and the principles of European Union law, the assumption, regulation and management of local public services of general economic interest are inspired by principles of efficiency in management, efficiency in satisfying the needs of citizens, production of quantitative and qualitatively adequate services, application of standard cost-oriented tariffs, promotion of investments in technological innovation, competition in the provision of services, subsidiarity, including horizontal, and transparency "(paragraph 1); "Promoting competition, freedom of establishment and the freedom to provide services of all economic operators involved in the management of local public services of general economic interest" (paragraph 2). The following general principles are also stated: "The administrations and bodies responsible for the organization of local public-sector services of general economic interest respect equal treatment, including between public and private undertakings, and operate in accordance with the principle of proportionality, Limits of what is necessary to meet the needs of the public interest "(paragraph 4) ..." Users of local public services of general economic interest are assured of accessibility, continuity, non-discrimination and the highest levels of quality and safety in in accordance with Article 117, second paragraph, (e) and (m) of the Constitution "(paragraph 5); ... "The parties responsible for the organization of local public services of general economic interest within the meaning of Article 117, second paragraph, (e), (m) and (s) of the Constitution shall ensure adequate forms of supervision and control and appropriate levels of user protection, according to the principles of subsidiarity, proportionality and fair cooperation "(paragraph 7)."

Midiri (2014) emphasizes the value of the competitive rules (community derivation) in the local public service regime and, in this regard, he considers that in the Consolidated Text there are goals and principles of management that are not placed in the exclusive interest of the users, but as a guarantee of the competitive dialectics. The users, indirectly, can obtain better service level as a consequence of the purposes of managing and increasing efficiency, lower tariff rates, promoting investment in technological innovation, increasing competition in the provision of services, promoting freedom of competition, freedom of establishment, and freedom of service of all the economic operators concerned. With regard to principles, however, one should consider equal treatment, including relationship between public and private, and this could mean a modern way to protect the private economic initiative.

The Disciplinary Status of Local Non-Economic Public Services

Following this idea of local economic services, also including non-economic or market-based services, which would be treated as economic services *tout- court*, would impose on the administration, under any assumption and guarantee of activities of general interest, deciding as from the Unique Text which are the "productions and supply of goods or services which would not be carried out by the market without public intervention or would be carried out under different conditions in terms of physical and economic accessibility, continuity, non-discrimination, quality and safety", and this would require to apply rigid organizational and functional models. This seems inadequate, says Midiri (2017) on the functional level, as well as disrespectful of the regional prerogatives enshrined in the constitution in determining non-economic and social services management models:

It is therefore necessary to identify the perimeter of local public services of economic interest drawn up by the Italian legislator on the basis of European law. In this way it will be possible, as anticipated, to identify the figure of local non-economic public services and their fundamental disciplinary status. As can be seen in the Single Text on Local Public Services, as in the case of public companies, there is no definition of non-economic services. However, it is possible to derive the notion of services of no economic interest from the notion of services of general economic interest in the local area and to specify the discipline that can derive from it. Also, on the occasion, it will be possible to check whether the Unique Text puts defined and unmistakable boundaries between the types of services that the administration entrusts to them or whether (as in the previous discipline) remains the discretion of the administration itself to determine whether, in own territory, an activity is likely to be carried out with economic / business or merely as a supply of service from public sector.

Derogations from the Community case-law on the competitive arrangements for service of general economic interest is the basis of the principles of subsidiarity and solidarity

In order to understand which kind of services are economic and which are non-economic ones, Ione should define a way to clearly understand when local services are not, even abstractly, susceptible of being subjected to market logic or being provided by market players. This way, they can be defined under the label of non-economic services.

Midiri (2017) points out that "the selection criterion appears to be drawn from the indications of the most recent Community case law on national derogations from the application of competition for services of general economic interest. In other words, the Court of Justice's indications as to the conditions for "subtracting" an economic service to competition may be valued in a different context in order to determine the cases in which the economic viability of the a service of general interest. It should be recalled here that the European order has always admitted national legislative derogations to the competition regime which affects services of general economic interest. According to settled case-law, "internal rules such as limiting the fundamental freedoms guaranteed by the Treaty may be justified if such legislation is adequate to ensure the attainment of the objectives pursued and does not go beyond what is necessary for that purpose." For example, "measures actually designed to provide health services (such as healthcare services) provided on behalf of public administrations to all citizens are reliable and of good quality, while minimizing public expenditure, are in principally capable of justifying a restriction on those freedoms "35. In essence, derogations are strictly proportionate to the purpose, if the market can not achieve the same ends as the general interest of the service or can not be guaranteed the basic conditions for its realization, ie effectiveness and financial sustainability."

Midiri also adds that the Court of Justice (2014) has recently confirmed this line of argument by stating that "Art. 49 TFEU and 56 TFEU do not preclude national legislation which provides that the provision of urgent and emergency healthcare services should be prioritized and with direct reliance, in the absence of any advertising, on voluntary partnerships, provided that 'The legal and conventional context in which the activities of the associations concerned take place actually contributes to the social purpose and the pursuit of the objectives of solidarity and budgetary efficiency on which that framework is based'. Thus, the need to fully realize the social ends of the service, the value of social solidarity and the financial sustainability of tasks of general interest are identified by case law as the real legal prerequisites for excluding competing rules for services with regard to to which there is a real or potential market. And again The Court of Justice, in its judgment no. 113 of 2014 states that "Articles 49 TFEU and 56 TFEU must be interpreted as not precluding national legislation which provides that the provision of emergency and emergency health services must be given priority and with direct reliance, in the absence of any advertising, on voluntary partnerships provided that the legal and conventional scope of the activities of the associations concerned actually contributes to the social purpose and pursuit of the objectives of solidarity and budgetary efficiency which is based on this rule "37. In essence, the European Court values, as prerequisites for the exemption from competition, the effectiveness of the benefits for the purposes of public interest, the balance of the budget and, above all, the realization of the value of solidarity (always in accordance with the principle of proportionality of national legislative measures).

The Principle of Horizontal Subsidiarity

The most important element, however, is that the Court in the same judgment no. 113, which concerned a law in the Liguria region, considered the principle of solidarity 38 as "assimilable to our constitutional and statutory provisions relating to the voluntary activity of citizens" and, more precisely, Article 118, last paragraph "of the Constitution, which provides that citizens" as individuals or associates participate in the pursuit of activities of general interest with the support of public authorities, on the basis of the subsidiarity principle ". The very core of the principle of solidarity is thus identified in the voluntary activity of citizens who carry out tasks of general interest for their community, which is recognized at the constitutional level as a fundamental content of the principle of horizontal subsidiarity."

This means that the Court recognizes in mutual help, or horizontal subsidiarity, a manifestation of the principle of solidarity which, in Europe, can be a limit to the application of competition rules. This is an innovative legal principle, as it opens a way for local public bodies to directly support local bodies which work on the territory, given that they don't provide a top-down service but propose a community based service in which the aspect of mutual help is central.

Starting from this, Midiri defines the principle of horizontal subsidiarity as a criterion for identifying non-economic services and their scope of activity:

"Given these considerations, non-economic services can well be identified in activities where the essential content of horizontal subsidiarity is achieved, such as the Italian implementation of Community solidarity.

These are, in particular, forms of participation in the provision of services by private individuals other than those in which the "social" intentions of the operators alone cannot neutralize the market logic in an objective way, closing it to other operators. This may be the case where the private operator runs an activity which is not likely to be absorbed by market logic because the specificity / subjective identity of the payee corresponds to a personal interest of the user or, in other words, to an essential feature for guaranteeing the concrete promotional effectiveness of the business.

At first the services covered by the self-help model, valued by the same law no. 328 and by regional lawmakers, fall into this category. This model is only apparently limited because the subjective homogeneity in the broad sense between the distributor and the service provider can also be implemented for various forms of community and associative structures. For example, co-operative communities made up of parents to help their children with disabilities or social marginality and then called by local authorities to open their activities to the generality of citizens or even communities of religiously characterized operators.

In this respect, it may be justified to exclude or mitigate the competitive awarding mechanisms provided for in the Single Text for structures thus achieving the services. The regional legislature could therefore adequately provide mechanisms for the award of services that value the principle of horizontal subsidiarity by "subsiding" the forms of delivery of services characterized by subjective homogeneity or "convergence" between social formations that perform the performance and the users.

The same considerations might apply to the exclusion of the broader protection mechanisms (such as the specification of the content of service contracts) provided by the legislator. With regard to the same forms of service provision, it could be useful to lighten or deform the relationship of protection between "close" subjects, in all cases where it is not useful to weigh the reports because it is the same "proximity" the use of complex instruments of protection.

All of which, of course, while maintaining the respect for the fundamental principles of public service and providing for other appropriate forms of guarantee of performance standards, including by regional provisions.

In general, submit the considered forms of provision of services to the indications of the law n. 328 and the regional implementation laws (which essentially reproduce the constitutional guidelines) finds its justification in that the form of delivery is characterized, as we have seen, by the closeness between distributors and users and is a peculiar expression of the principle of subsidiarity. This should allow the regional legislator greater margins of maneuver to guarantee the implementation of this principle.

For social services, however, those of general economic or market interest, the rules of the foreseen Unique Text, which specifically aim at harmonizing social ends with the protection of competition, seem to prevail over the general indications of law no. 328 and constrain regional legislative autonomy."

At this point, it must be considered (Midiri, 2017) that the notion of service of non-economic interest should essentially have two characteristics. The first one is that not even after a subjective assessment of the administration it can be evaluated as capable to enter the market, because this can not even manifest itself in potential as a market activity. The second one is that it's an activity that has characters of public interest that require the public infrastructure to ensure its realization.

On the basis of these definitive coordinates, the new text of the law erodes the scope of non-economic services of general interest and thereby reduces the scope of activities of public interest deriving from competition and the national and European legal system in this field.

On one side the legislature, by introducing a broader concept of public service of economic interest, and restricted to non-economic public services, ensures the respect of the competition regime and excludes all derogatory legislative interventions aimed to extend the range of non-economic services through flexible national laws. On the other side, defines the margin of choice of the local authority as stated in the Unique Text: "in the persistent absence of a general discipline on the regime of activities and the reliance of local public services", p. 404.

An important point is that in this legal framework, services of non-economic interest are not identified with the social services, as also proposed during the "*Prenez la parole*" 2013 European Conference on Social Entrepreneurship, as some of them are located in a market that is at least potential and therefore fall into the category of services of economic interest. As a consequence, Midiri (2017) says "in the field of non-economic services, which cannot be located in a market context, there may be places of public interest that are not properly qualified in terms of services aimed at overcoming situations of people's need, or in terms of of social services in the classical sense. Reference is made to the combined provisions of Art. 128 of Legislative Decree no. For the purposes of this legislative decree, "social services" means all activities relating to the provision and delivery of free and paid services or economic benefits designed to remove and overcome the needs and difficulties of the human person encounters during his / her life, excluding only those insured by the social security system and the health care system."

Midiri then reminds the Multifactories, "where economic-social development activities are carried out, which do not result in the provision of contestable services on the market."

The outcome is that to a lesser extent of the private providers of social services will be applicable the scheme laid down by law 8 November 2000 no. 32831 and the state and regional implementing rules (such as March 30, 2001, which the regions tend to apply). More precisely, a greater share of social services takes on the features of services of general economic interest. They are therefore attracted to the competitive regime laid down in the Consolidated Law and are excluded from the uncompetitive derogating regimes provided by yielding state legislation and regional legislation.

As is well known, currently (besides direct public management) the involvement of private individuals represented by enterprises, entities (with various forms of recently-reorganized non-profit organizations), families and individuals, is based on trust mechanisms in the case of the participation of the providers in the system) disciplined, precisely, by the dpcm n. 31 of 2001, by adapting public evidence according to specific criteria (such mechanisms in some regions assume, for example, accreditation)

The notion is also different from what appears to emerge from the Single Text on the Third Sector, the provisions of which should therefore be coordinated with the rules in question.

The above discussion concerns European legislation, but is generally valid as a conceptual reflection. Therefore, it is clear that the type of service excluded by the legislator from the spectrum of competition on the market concerns only those services that are provided with a view to mutual aid and that can also be theoretically excluded from competition on the free market because they are unfeasible as a service released from the mutual relationship. In practice, it is a relationship of mutual aid that prevails over the type of service and also over the person who puts it into practice. It is quite clear that this mutualistic relationship between companies can be fully realized within a context of community economy, and specifically within shared workspaces, when and if there is an effective mutualistic relationship among members.

CONCLUSION

The businesses led by crowdworkers are a hybrid of craftsmanship, free profession and entrepreneurship, but usually take on many of the traits of craftsmanship when it comes to production methods and also follow many of the typical ways of artisans to relate to the market and to deal with large capitalist companies. Crowdworkers, like craftsmen, are subjects operating on the free market, but they are not capitalistic operators, because they lack the typical features of capitalist companies. In particular, one of the founding values of crowdworkers is the sharing of knowledge and means of production, which is opposed to the competitiveness typical of the capitalist system.

Therefore, the concept of exchange is fundamental, so it is useful to compare the mutual aid model of shared workspaces with the sharing economy model.

In particular, it is necessary to evaluate the different role of the concepts of "exchange" and "community", as they are interpreted by the operators of the sharing economy and within shared workspaces.

In the first case, the focal point is the exchange itself, while the community is merely an element of support for the exchange action. In the second case, it is the community that is at the centre, and the exchange is a consequence of the trust that is created between the members of a community. It is clear, therefore, that the real point that differentiates the two approaches lies in the role of exchange and community. In the sharing economy the main focus is on the very act of exchanging a good, or making it available, and the community is somehow accessory. This is what happens with Uber, for example, where the community is called upon to give a rating of the actors involved, but it is an ex post role, that is, following the act of that economic system, a race in exchange for a payment. An essential element for having a contact between two elements of the community, therefore, is not belonging to the community, but the performance of an exchange.

In the system of shared workspaces, on the other hand, community is the pre-existing and fundamental element so that exchanges can even be hypothesised. To some extent, the community itself regulates the nature and possibility of exchanges, which are an event that follows the formation of the community. Therefore, exchange is something that happens in communities, but the relationships between subjects belonging to the community can exist even without a direct exchange.

Therefore, it seems much more coherent to speak, in reference to shared workspaces, of community economy, a term that better represents the central role that the community plays.

The role of the community is also central in determining whether or not the mutual aid activities of shared workspaces are conducted in compliance with competition law and the rules of the free market, or whether it is a form of unfair competition towards a commercial offer otherwise reserved for professional operators.

The answer lies in the concept of mutual help, which is closely linked to the relationship of trust that takes place within a shared workspace. This report makes the exchange of aid a non-economic service, which also takes on the value of general interest as it promotes the growth of the competitiveness of the entire territorial system.

These are therefore economic operators who relate to the free market, but who base their actions on a concept of competitiveness based on a relationship of mutual aid, possible only in the presence of a deep trust that only a community project can guarantee.

These economic subjects, the crowdmakers, can therefore effectively make the fact of being part of the community a competitive asset of their own, proposing a viable alternative to the capitalist competitive model, but without going beyond the context of free competition on the free market.

REFERENCES

Asher-Schapiro, A. (2014, September 19). Against Sharing. *Jacobinmag*, 1. Retrieved from: https://www.jacobinmag.com/2014/09/against-sharing/

Bernhardt, A. (2014). Labor Standards and the Reorganization of Work: Gaps in Data and Research. *IRLE Working Paper*, *100*(14). Retrieved from http://irle.berkeley.edu/workingpapers/106-14.pdf

Brescia, R. (2016). Regulating the sharing economy: New and old insights into an oversight regime for the peer-to-peer economy. *Nebraska Law Review*, *95*, 88–144.

Calloway, A. (2016) Apploitation in a city of instaserfs: How the "sharing economy" has turned San Francisco into a dystopia for the working class. *The Monitor*. Retrieved from: https://www.policyalternatives.ca/publications/monitor/monitor- januaryfebruary-2016/

Cherry, M. A. (2016). Beyond misclassification: The digital transformation of work. *Comparative Labor Law & Policy Journal*, *37*(3), 544–577.

Commission Communication of 26 April 2006 [COM (2006) 177], Social Services of General Interest in the European Union, para. 2.1 & 2.2.

Comunicazione della Commissione di data 26 settembre 1996 *I servizi d'interesse generale in Europa* [COM(1996) 443] in *Gazzetta ufficiale* C 281 del 26.09.1996 punto 18.

Council of State. (2013). Sect. III, 15/04/2013, n. 2056. *CDS*, *4*, 911.

Court of Justice. (2015). Sect. V, 11/12/2014, n. 113. *Pharmaceutical Law Review*, *1*, 198.

Focardi, G., & Salati, L. (2015). A New Approach to Knowledge Sharing, the Multifactory Model. In O. Teràn & J. Aguilar (Eds.), *Social Benefits of Freely Accessible Technologies and Knowledge Resources*. Hershey, PA: IGI Global. doi:10.4018/978-1-4666-8336-5.ch009

Friedman, G. (2014). Workers without employers: Shadow corporations and the rise of the gig economy. *Review of Keynesian Economics*, *2*(2), 171–188. doi:10.4337/roke.2014.02.03

Gallo, D. (2011). *I servizi di interesse economico generale, Stato, mercato e welfare nel diritto dell'Unione Europea*. Busto Arsizio, Italy: Giuffre.

Hamari, J., Sjöklint, M., & Ukkonen, A. (2015). The sharing economy: Why people participate in collaborative consumption. *Journal of the Association for Information Science and Technology*. Retrieved from https://ssrn.com/abstract=2271971

Konig, R. (1968). *The community* (E. Fitzgerald, Trans.). London, UK: Routledge & Kegan Paul.

Koopman, C., Mitchell, M., & Thierer, A. (2015). The Sharing Economy and Consumer Protection Regulation: The Case for Policy Change. *J. Bus. Entrepreneurship*, *L*, 529.

Kostakis, V., & Bauwens, M. (2014). *Network Society and Future Scenarios for a Collaborative Economy*. Basingstoke, UK: Palgrave. doi:10.1057/9781137406897

Langley, P., & Leyshon, A. (2016). Platform capitalism: The intermediation and capitalisation of digital economic circulation. *Finance and Society*, *1*(21).

Lobo, S. (2014, September 3). *Auf dem Weg in die Dumpinghölle*. Spiegel Online. Retrieved from: <http://www.spiegel.de/netzwelt/netzpolitik/sascha-lobo-sharing-economy-wie-bei-uber-ist- plattform-kapitalismus-a-989584.html/>

Lubow, L. B. (1987). *Artisans in Transition: Early Capitalist Development and the Carpenters of Boston, 1787-1837*. Los Angeles, CA: University of California Press.

McMillan, D. W. (1976). *Sense of community: an attempt at definition* (Unpublished manuscript). George Peabody College for Teachers, Nashville, TN.

McMillan, D. W. (1996). Sense of community. *Journal of Community Psychology*, *24*(4), 315–325. doi:10.1002/(SICI)1520-6629(199610)24:4<315::AID-JCOP23.0.CO;2-T

McMillan, D. W., & Chavis, D. M. (1986). Sense of community: A definition and theory. *American Journal of Community Psychology*, *14*(1), 6–23. doi:10.1002/1520-6629(198601)14:1<6::AID-JCOP22901401033.0.CO;2-I

Midiri, F. (2014.) Promotion of competition and judicial union: the events of local public services. *Quarterly Public Law*, *133*.

Natour, F. (2016). Respecting Human Rights in the On-Demand Economy: Closing the New Governance Gap. *Business and Human Rights Journal*, *1*(2), 315–320. doi:10.1017/bhj.2016.7

Opinion of the Council of State. Opinion No. 01075/2016 dated 03/05/2016, p. 13 and 14, but cf. also p. 15.

Pumpipumpe. (n.d.). Retrieved from http://www.pumpipumpe.ch/the-association/

Stott, R. (1996). Artisans and Capitalist Development. *Journal of the Early Republic*, *16*(2), 257–271. doi:10.2307/3124249

Taylor, G. R. (1951). *The transportation revolution, 1815-1860*. New York: Rinehart.

Yaraghi, N., & Ravi, S. (2017). *The Current And Future State Of the Sharing Economy*. Brookings India.

Chapter 5
Shared Workspaces Presented by the Communities

ABSTRACT

In this chapter, the authors present some relevant examples of shared workplaces they visited during the research. These are not to be intended as best practices. They are not examples of multifactories. These places are important because in each of them the authors found some relevant inspiration, issue, or practice, in order to develop the multifactory model. Authors visited and studied all of these places during the first step of the research while building the multifactory model (2012-2014) or the second step, the testing phase of the model (2015-2016). For this book, the authors asked some spaces to describe themselves following a similar presentation path. The authors wanted them to explain their own specialities and their motivation in order to better understand similarities and differences. The following chapter is based on these personal communications, which have been slightly modified in order to better stress specific points.

INTRODUCTION

As seen above, there are many different reference models to develop a shared workspace and, leaving aside coworking spaces, for what concerns comanufacturing spaces there are many models that can hybridize and take on different nuances depending on the application context. Yet each space is identified with an idea and transmits it both in the organizational model and

DOI: 10.4018/978-1-5225-7089-9.ch005

when it presents itself. For this reason, authors find it interesting to propose in this chapter some examples of self-presentation of various comanufacturing spaces, to highlight the differences through their self-representations.

The Authors agree with Senge when he says he believes that "benchmarking best practices can open people's eyes as to what is possible, but it can also do more harm than good, leading to piecemeal copying and playing catch-up (Senge, 1990). As one seasoned Toyota manager commented after hosting over a hundred tours for visiting executives, "They always say 'Oh yes, you have a Kan-Ban system, we do also. You have quality circles, we do also. Your people fill out standard work descriptions, ours do also.' They all see the parts and have copied the parts. What they do not see is the way all the parts work together." I do not believe great organizations have ever been built by trying to emulate another, any more than individual greatness is achieved by trying to copy another great person." The spaces presented are therefore very different from each other and are proposed in this chapter not to serve as inspiration, but rather to help form a vision as broad as possible on shared workspaces and specifically on the experiences of comanufacturing.

Building Bloqs in London was born from a social entrepreneurial drive of a group of craftsmen and designers and has developed over the years as a maker space. Freiland, from Potsdam, was born as a youth cultural space by a group of young people driven mainly by political motivations and then turns into a multifunctional place that hosts several craftsmen and a fab lab. In Dresden Konglomerat, it was born as a place of aggregation of neighborhoods, it develops as a travelling fab lab and becomes in the end a space dedicated to digital technologies and making. The Munich fab lab is a large and structured fab lab, managed by an association and designed to offer a place to members where they can develop their projects, mainly in their free time. OZ Officine Zero is a squatted space in Rome that was created inside a factory once used for the maintenance of railway sleeping cars. Following the bankruptcy, the workers occupied the factory workshops to start micro craft activities, and later joined other professionals and craftsmen creating a project of comanufacturing that between 2017 and 2018 has started a process of regularization together with the city institutions and the ownership of space.

BUILDING BLOQS: LONDON

This section is based on personal communications that took place between September and October 2017 with Andrew Catcheside, Arnaud Nichols, Al Parra, Vinny Nanray founders of Building BloQs.

Introduction

Building BloQs represents a new approach to space, resource, and community for London: a progressive solution for freelance makers and small manufacturers to counter the rising costs of the city.

Opened in 2012, Building BloQs is a not-for-profit open workshop created so that London's makers, small businesses and designers can access tools and space for work, but also the extra resources they need to grow their careers.

With a huge range of professional-standard equipment in its north London workshop, anyone who wants to work in wood, metal, textiles, paint, and digital fabrication can do so. Building BloQs operates a flexible pay-as-you-go membership model so its members can scale their use of space and equipment up, down and sideways to fit their needs and budget.

As a result of its equipment spec and fair terms, Building BloQs has quickly become the largest and best equipped workshop of its kind in London. But the facility's magnetism is also down to the emphasis it places on nurturing and promoting its membership: a diverse community of designers and makers that numbers over 350.

Made at BloQs and the BloQs Story

In its latest initiative to galvanise the talents of the community, Building BloQs has launched its Made At BloQs service - a design, fabrication, project management and install service – to open new revenue streams for the facility and its membership.

Made At BloQs launched in 2017, five years – almost to the day – since a group of soon-to-be-homeless housemates signed a lease on an empty, leaky building in an industrial pocket of Enfield, north London.

The housemates were a mix of makers, craftsmen and artists who had spent several thrifty, happy years cohabiting in a north London live-work space; sheltering from rising city rents. In 2012, their landlord served notice and the reality of the market hit home as they scouted for new premises.

When one of the group spotted the aforementioned leaky building, the group saw a viable space in which they could create a new kind of workshop model.

Over 2013, co-founders, Arnaud Nichols, Al Parra, Alex Motta, Vinny Nanray and Julien Thomasset, ploughed in their own cash and labour and a vision started coming to life. It would be a workspace shared by many: people would pay for it as and when they used it. It's the same today.

Building BloQs works via simple economies of scale: more members equals more purchase power. The more cutting-edge machines and equipment that come in, the more people use it and the more membership grows.

With solid supplier connections and friends in the local trade community, Building BloQs is able to pass material savings onto members. Saving costs is always the name of the game - leftover waste materials, for example, are recycled and upcycled between members. Or converted to fuel to power the building.

To cater for as many makers, disciplines and talents as it can, Building BloQs combines best-in-class machinery – from traditional analogue tools to advanced digital systems – with a unique social ecosystem. With that, the facility has evolved into a happy, collaborative, progressive place to work.

The social hub of Building BloQs is an on-site cafe which serves artisanal food for knock-down prices. The cafe is a focal point for bringing members together - introducing makers to one another and nurturing a culture of skill-swapping and co-working. Building BloQs proudly fuses people and skills together so that its members can bid for bigger jobs. As of 2017, many new businesses and partnerships have been matchmade at BloQs.

The progressive environment at Building BloQs continues via its training initiatives. Co-founder Vinny Ninray coordinates on-site training which often means pairing members up so they can trade skills. At BloQs, it's not uncommon to see blacksmiths working with carpenters and seamstresses with digital fabricators. By cross-pollinating skills, makers are able to diversify their offerings in the market. And with a wider range of services to one's name, earning power can grow.

Building BloQs has come a long way since that group of friends stood in that empty shed. The building, the faces, the facilities have all changed – but the mission has not.

The Mission

Even for the co-founders, it's almost a shame that Building BloQs has to exist. Making and manufacturing both have a strong legacy in UK cities. But with new-day urban phenomena such as rising costs and gentrification, makers are being priced out of town - abandoned in their hour of need.

In summer 2017, co-founder Al Parra was invited to address the London Real Estate forum. He had a platform to urge that City Hall do more to more support makers. And in his keynote speech he presented Building BloQs as a viable solution for the long term.

Prevalent in Al's speech was the eroneous idea that facilities like Building BloQs – and other open workshop projects – are some kind of 'meanwhile' solution, a holding pen while the city figures out how to manage making and manufacturing.

For Al and his co-founders, Building BloQs is the lasting solution. The organisation and its business model has proven itself in real terms, in real time, with real people.

"We aren't about meanwhile," Al said. "Workshop spaces need to be removed from that kind of thinking. There are solutions on the ground now. There are people working in our workshops voting with healthy P&L sheets.

"Building BloQs is a new kind of factory for a new day, it's a space for innovation and to help usher in the city's next phase. We're growing quickly and we're growing organically."

The message of Building BloQs is that cities, like London, once again need to work for their makers. The city needs workshops – and a workshop model – that provides adequate space and resource on a new plain to help surface untapped maker talent. It needs a new way of facilitating and promoting the collaboration and the innovation on which the future will be built.

The Building BloQs solution is not born purely of nostalgia, altruism or loyalty. The economic results are robust.

Makers at Building BloQs make a solid living. Not only do they they pump revenues back into supply and service in the local area, their incomes go towards tax, and spending in local communities. Makers who work at Building BloQs earn anything between £20,000 and £120,000 per annum - and with a growing membership, those are numbers the city needs to keep hold of.

The Next Phase

If the current site has proved the viability of the model, the next logical step is to grow things. With that, Building BloQs is set to move to a new location several times the size of the current spot.

A short distance from home is a new site which measures 40,000 square feet. This is Building BloQs' next stage: it will enable the business to fully investigate the possibilities of scale, and explore the savings and the benefits that come with a bigger and better facility.

With yet more purchase power, clout and more talent across more disciplines, Building BloQs aims to enable, drive and develop the skills and careers of makers, creatives and thinkers of many more trades and disciplines than it currently offers.

It Nearly Didn't Happen

Building BloQs has a range of fans and admirers - in industry, in government and in the press. Not long ago, Aditya Chakrabortty, senior economics editor at UK newspaper the Guardian, sent out an SOS message when the facility was in jeopardy (Chakrabortty, 2016).

Aditya's red flag was instrumental in taking Building BloQs from the point of collapse and pushing it straight to the forefront of the urban planning agenda.

"Everyone who came across Building Bloqs loved it – loved what it symbolised and the young men who'd charged into setting it up. Yet when I first dropped by, it was close to shutting down. Its status as a new social enterprise – in which any profits were reinvested back into the business – meant that the banks wouldn't lend and the venture capitalists wouldn't invest. The novelty of a makerspace in their backyard – freelancers in metal and wood and plastic hiring benches by the day and sharing workshop kit and knowhow – confused Enfield council, who couldn't work out how best to help.

The roof had a giant gash, the place was freezing, the directors were working without pay till 10 at night and then crashing on friends' sofas. I had a row with a well-intentioned local councillor over what else he and his colleagues should be doing, filed a column that read more like a distress flare, and went away fearing the worst.

Well, as they say on Facebook posts, you won't believe what happened next. First, mail began pouring in from Guardian readers asking how they could help. Some offered loans. One couple just gave £5,000.

"That made the difference between Building Bloqs staying open or not," remembers Al Parra, a co-director. More politicians popped in, including the Green party leader, Natalie Bennett, and council officials sent over details of available grants. We (were later) awarded £2.7 million of support from the Mayor's London Regeneration Fund (LRF)."

With the mission standing firm, and the business of BloQs settling, makers from all backgrounds continue to call on the facility and its evolving spread of services.

When an on-site textile studio was opened in 2017, students from London's fashion universities found a new creative home.

Mature student Zana Agafonva designed and created her final collection here. After graduating in June, she's proud to still call Building BloQs her professional home.

Zana Case Study

Zana Agafonva arrived at Building BloQs in spring 2017, ahead of her graduate show at The Cass fashion school at London's Metropolitan University.

In a short space of time, her awe-inspiring collection of leather-based fashion came together and Zana has become part of the fabric at BloQs: an inspiration in the textile studio, and a magnetic addition to our community.

After a career in fashion that started in her native Lithuania, then part of the Soviet Union, Zana moved to London. Although she had a considerable professional portfolio, she was a latecomer to the English language and struggled to find work in the capital.

After studying an access course in fashion at a local college, Zana enrolled at The Cass to complete a BA (Hons) in Fashion Design. And at the beginning of 2017, she started searching for an alternative workshop, away from the oversubscribed studio facilities at her university.

"I was looking for somewhere I could be. And think. And have the facilities to make a large collection in a short space of time. After looking around, I found Building BloQs," said Zana.

"At first I thought I could buy a machine and do it at home but I didn't have space. So I phoned a few studios but one was too small; one had only a domestic sewing machine; one wanted me to bring in my own sewing machine; others were fully booked.

"And then I came here. The space is full of creative people which is really nice. People are friendly and happy to help me. It's fun. It's a creative environment and people like to see what other people are doing and get involved.

"There's good machines, including a walking foot sewing machine which is very important. And space. There's a lot of space."

In just a few short weeks, Zana put the facilities at BloQs to use and compiled an entire 11-piece collection of women's fashion made almost exclusively of leather.

"Salvador Dali always captures the kind of emotion that I want to put across. I've made very strong pieces but there's a playfulness and a lightness to them too. There is a positivity there."

Like most students, Zana isn't sure what the future holds beyond her studies but she is planning for two things: one, that she can, once again, run her own business. And two, that she can continue to work out of Building BloQs.

"In here I don't need to bring machines, I don't need more space. I don't have to be here every day," she says. "I can come only when I need to - and I can pay only when I need it.

"When I'm ready, I know Building BloQs will be ready."

The Importance for Multifactory Model

Building Bloqs was the first space the Authors visited outside Italy. At the time of the first visit, in February 2013, Building Bloqs was in its very early stage and it was possible to see what happens at the beginning of the development. The first important element was that to build a community-based project, the managing entity should be a nonprofit organization, or a social company. Private companies can manage top-down organized spaces like coworking spaces, but don't fit the needs of community-based projects. Another important issue, linked to the first one, was the way to structure the monthly fee participants pay to the project. This fee should be flat, all-inclusive, and without any barter between efforts to develop the community and the monthly fee.

FREILAND POTSDAM

This section is based on personal communications that took place in 2015 with Achim Trautvetter and upon the document "Betreibkonzept FreiLand" distributed by Cultus UG in 2016.

FreiLand is a Cultural Center in Potsdam (DE) in a 12,000 square meters area formerly owned by the local gas and water company. It is made up by five different buildings and a huge outdoor area.

By 2017 it had become a multi-purpose cultural space, home to more than forty projects, starting from a Night Club called Spartacus, a bar-café Haus Zwei, different seminar rooms, a youth club Mitte, the Frrapó free radio, some rehearsal rooms, an open studio for artists Laboratorium, a theatre room, a gym, different offices and studios. The size of the outdoor area and the six buildings offer particularly good conditions for a youth center in a rather central location. It is surrounded by commercial activities and has good transport facilities, it borders with only a few residential buildings. The area is very easy to get to by bicycle and it is a ten-minute walking distance from the Potsdam main train station.

The Story

FreiLand-Potsdam was born in 2012 as a result of a political agreement between a conspicuous local group of left wing citizens and the public institutions. The Potsdam municipality gave the opportunity to turn an empty space into a functioning cultural center with no rent costs and granting the project 125.000 euro a year for a testing phase of three years. The management of the overall space was officially entrusted to CULTUS UG, a company with no profit purposes founded specifically for operating as a Youth Culture center with the purpose of the promotion of arts and culture, political education, child and youth welfare.

In the concept proposed for the testing phase, FreiLand was conceived as an area for a self-determined and self-organized youth culture, in particular for youngsters aged between 12 and 27 years. CULTUS could build good connections with Stadtwerke Potsdam and the municipality in itself. During the testing phase, FreiLand developed into a lively, participatory and pulsating cultural place. In 2015 FreiLand had successfully mastered its testing phase, receiving a positive evaluation from the institutions and received other ten years of contract for the same space.

Some important results out of the testing phase:

- The first few years have shown that the age span represents the vast majority of users, especially the Spartacus Club, managed by young people, located in the center of FreiLand. However, the restriction to a target group between 12 and 27 years did not fully meet with the reality.

- FreiLand hosts many projects which are intergenerational and also explicitly address people of all ages. FreiLand has developed into a "free land", a space open to everybody, insofar as they do not restrict other people through racist, violent, homophobic, sexist or otherwise discriminatory behavior.

Organization

FreiLand is characterized by a governance of horizontal participation, combined with an organization that gives autonomy to individual projects to be developed as they wish, in which mutual respect, democracy and self-determined action are part of everyday life. There is not a bottom down approach but users and activists design and shape FreiLand itself bringing their own projects into life inside FreiLand. The barrier-free, pulsating and lively character of FreiLand has been seen by users and visitors as an opportunity to experiment and develop and test their own ideas.

This result is achieved by the special profile of the overall project.

CULTUS, together with the groups hosted inside FreiLand has developed the current governance system;

The main goal was to integrate all user groups into processes in which they are able to:

- Make decisions together on the use and development of the overall project.
- Develop content and future concepts and plan their implementation.
- Openly discuss perceived hierarchical structures inside the community.
- Provide an equal treatment and a transparent decision-making in the governance system.

The main role of CULTUS is the provision of free spaces and the development of possibilities. It arranges spaces, situations and opportunities for users in which they can autonomously and flexibly shape their programmatic contents. Its main task is to actively support the projects without having to interfere with the content.

CULTUS focuses on the following tasks:

- Support, coordination and monitoring of the FreiLand users (Spartacus e.V., bands, sprayers, theaters, sports facilities, studios, seminar and education areas, etc.).
- Accompany and coordinate the users' own efforts as well as the maintenance activities.
- Support the networking of the projects with each other in order to focus and develop synergies and co-operation.
- Integration of the Street Art scene into the overall project and the associated artistic-conceptual development of the walls with the protagonists of the Street Art movement.
- Support the outward-looking FreiLand visual identity in consultation with the FreiLand members, which includes diverse media (such as facebook, google +, blogs, etc.).
- Support for the inward communication channels such as mail distribution, forum and facebook groups.
- Transparent and open management (regular report to the user plenum of FreiLand).
- Expansion of an accommodation area together with the Spartacus e.V. and the seminar area responsible for the accommodation of artists and seminar participants.
- Development of the café, Haus 2, as a meeting place and meeting space.
- Organization of the seminar rooms for external users as well as for internal cultural and political projects.

The CULTUS deliberately decided to also become an event organizer. The cultural-artistic contents of the social areas are decided by the community aside from two to three big cultural events per year, where the CULTUS takes the main organizational responsibility.

The continuous presence of new honorary volunteers as supporters, mostly coming from the Civil Service program, also ensures that the project receives constant help, support and feedbacks from younger generations. During the testing phase, it was evident that there was a strong need for specialized technical support for the maintenance of the buildings and the green area infrastructure. The ordinary and extraordinary maintenance can't be organized solely on the basis of voluntary help by the community.

Space and Permanent Projects

Outdoor Area and Graffiti Walls

The huge green area around the five buildings allowed FreiLand to offer Self-organized second hand markets, workshops, educational and group offers, youth welfare programs, as well as small and large family festivals.

The Freigarten Potsdam crew, which has an office in Haus 1, is currently developing an urban gardening project involving all FreiLand members in some urban gardening actions. Every year, new green areas are being developed.

Nine hundred square meters have been aimed at legal spray art.The street art project has been shaping FreiLand since its early beginning. On the one side, there are areas which are constantly changing and are often used for learning and testing spray techniques by young writers. On the other side, some of the large walls on the site are designed as 'concept walls', which are maintained for a longer time. Every year, during the street art festival, almost every wall is painted with a new art piece that accompanies FreiLand for a whole year.

Haus 1: Office and Studio House

The entire building has a floor area of approximately 1200 squares meters divided in two floors. Haus 1 has been renovated and converted by the members, small rooms have been transformed into larger rooms without changing the overall building structure. The Haus 1 became home to offices and studio spaces, it is also a place for the CULTUS UG coordination office. On the first floor, many projects, freelancers, associations and organizations have found an office space. The seminar and theatre rehearsal room, which have been expanded during the testing phase, are a good stage for hosting opportunities for temporary education. The FreiLand plenum has also opted for another common space, the Community Kitchen Office. Projects and individuals who only need a temporary workplace can use it free of charge. Furthermore, the room includes a fully equipped kitchen and a large table, so that the users can meet here for a joint meal and have an informal exchange.

On the ground floor, one open area became a versatile atelier, with a large open studio available to artistic activities. This common atelier is very close to the Laboratorium, an atelier for free artistic expression self organized by a group of artists.

On the ground floor there also is a gym area, completely self-organized by different sport organizations: Aikido, Bauchanzanz, Bujinkan Budo Taijutsu, Capoeira GRAZ. They offer sport activities open to the public. Haus 1 is also home for the Office for Youth Participation in Brandenburg (JUBB) coordination office and the Food Chain Potsdam e.V., a cooperative of approximately seventy members that collectively order together fair, regional and ecologically sustainable products.

Haus 2: Café "Hauszwei"

A café in a large cultural center is also a place of exchange and encounter. The Freiland café "hauszwei" has become a popular hub and a popular venue for cultural formats such as readings, film and toddler activities. The CULTUS also tried to offer a regular lunch at affordable prices from its own resources and with the help of guest cooks, but the project wasn't financially sustainable and ended after few months.

In addition, the café became home of the free radio Potsdam – Frrappó that broadcasts live every Monday from the café, and once a month there is a Fifth Friday when it becomes a real radio show with live music and live interviews, while the public is very close to the action.

Haus 3: Spartacus Club: Central Event Building

The Spartacus in Haus 3 is the "heart" of FreiLand. The event building has been financed by the Sparkasse e.V. The organization is managed by around eighty young people. They organize everything from the event planning, to the financial accounting, gastronomy, admissions management. Haus 3 is independently managed by the Spartacus association but is also open to other interested parties who can ask for the use of the Spartacus club during the plenum meetings. The house has a floor space of about four hundred fifty squares meters and is designed for a visitor capacity of 350 people.

Haus 5: Workshop, Artist and Seminar Accommodation, Camp

It is intended to accommodate a wide range of uses. About 500 squares meters are storage areas because they were found in a rather good construction condition. Over the last few years a tech and science workshop, the MachBar, has been slowly developing: it consists of four rooms on the ground floor and has a total size of approximately one hundred squares meters. The MachBar concept: "The MachBar combines a traditionally handy workshop with a Fabrication Laboratory (FabLab). The goal is to represent societal issues such as re-cycling, up- and down-cycling, urban ecology, consumption, resource consumption, lifestyle and life forms, production processes, open source, open science, political participation.

The wood and metal workshop, during the first few years, was used mostly for repair and maintenance work for the all FreiLand common areas. On the first floor there is an area of about eighty squares meters which is to be expanded to a group accommodation for seminar and artist groups. This includes two large and a small bedrooms.

And to develop their own proposals and concepts, the workshop is a space in which people are able to deal with their own responsibility and self-determination with themes that affect themselves and their assigned role and function in society.

Everything collected from events room rental or other activities, between 320,000 and 330,000 euro/year, is reinvested into the project, to pay workers, to improve the infrastructure, to build a new cultural project to be offered to the citizens. Since the beginning FreiLand was designed with an extremely low staff quota: two cultural managers full time and three part time positions. Since the beginning it was clear that FreiLand can only function if members of the community are willing to take on personal tasks and responsibilities, voluntarily. This positive effect can only be achieved through the identification of the volunteer with the project. After a few years, when the café started to have make an income, the staff could be increased.

Spatial conditions of FreiLand will always change. In some areas of the building, the usage is defined, in other areas we adapt to current requirements.

The CULTUS sees its methodological approaches focusing primarily in offering free space, rooms to people to develop their ideas. The members plenum meets once a month and consists of representatives of each group.

The Plenum

During the members plenum the current situation of the entire project is communicated and shared, in terms of finances and current work in the everyday business. The plenum makes its decisions based on a democratic basis following the consensus methodology. Every new topic is discussed by the plenum and if there are voices against some proposals, these are registered and can monitor the evolution of the project.

It is important that the single projects coordinate themselves in their own separate plenum structure, as the volume of contents and number of people involved is exceeding the possibility to be managed in a single meeting.

These plenum structures are accompanied and supported by the sponsor, but they are intended to work independently from the sponsor and to organize their interests. The sponsor is in any case always available on an advisory and supportive role, if requested by the user groups.

The CULTUS has the overall responsibility for the site and is therefore invested in having a very good relationship with all members. An important part of Cultus is related to insurance and legislative issues. It provides knowledge in this area to the members and advises them, if it is within its possibilities.

The project is changing the size of FreiLand every day, and the important role of CULTUS is to let communication circulate, both internally between the different projects inside FreiLand, and externally towards all citizens, and information outlets, in addition to the regular monthly plenary session and the fixed meetings times.

CULTUS, in order to reach this goal, adopted some online tools and developed its own.

Tools for a Democratic Participation: DISCOURSE and the PROTOKOLL

Instead of using Google Drive, Dropbox or any other management tools available on the web, since 2015 FreiLand has been using an open source platform called *Discourse*.

It provides an online forum, the web platform Discourse, accessible to all members of FreiLand. There, it is possible to ask questions, to organize plenary sessions and events, read protocols, and users can also quickly communicate their current work situations there. In addition, the forum is organized in a way that allows the individual user groups to create(?) areas that can only be viewed and written on by the users belonging to one group. This access is extremely low barrier and allows everyone to participate in the decision-making processes outside of regular dates.

The Importance for Multifactory Model

FreiLand was very important to better understand what happens when a space is organized in multiple buildings, and especially the importance to clearly define functions and relationships between buildings starting from layout of the site and flows of people. The FreiLand experience also suggested the importance to avoid the coexistence of workers and associations, or people who just want a place for activities during spare time, as their needs, efforts and perspectives are completely different. Another important suggestion that came from the FreiLand experience was to keep a direct link with the local government and keep it always updated and aware of what happens within the space. An inspiring idea(?) in managing the community was to adopt the method of consensus when making decisions, and the use of the Protokol to prepare meetings and to keep everybody updated. The last important point was the importance to divide the managing entity of the space, which should be a not-for-profit entity like an association, and the entity responsible for providing services, managing a sustainability plan, supervising the project and managing financial issues, which can be a company, related to the space and to the managing entity by a clear agreement defining limits and intervention areas.

KONGLOMERAT IN ROSENWERK: DRESDEN

This paragraph is based on personal communications that took place between August and October 2017 with Bettina Weber, Ina Weise, Anna Seidel, Michael Hunstock members and founders of Konglomerat.

Today

The konglomerat e.V. is a registered non-profit association placed in Dresden, Germany. The over 370 m^2 industrial complex offers the current eighty members different workshops such as wood, metal, serigraphy, sewing, bike, plastic or printing. Furthermore, the FabLab provides different technologies as laser cutting, 3D printing or CNC milling. Alongside the workshop,s the association supports and invests in education, creativity and culture as well as charity. A wide and diverse group of people pays a small monthly service charge and gets an open 24/7 access to the community and the workshops.

Furthermore, interested individuals or groups can ask for organisational or legal support.

The organisation is based on community services. The five-headed and elected board occupies itself with different structural subjects e.g. accounting or community work. The different workshops or particular subjects organise themselves in an independent manner. Once a month, the whole association meets up to discuss and decide about current subjects.

The konglomerat e.V. takes in account the different interests of its community and gives room to different cultural, societal and manufactory-technical activities. Throughout active and collective creation of the environment, personal abilities can be developed and a positive impact to the surrounding can be created. This corresponds to the underlying philosophy of a collaborative and social culture: "Do it together". Like Joseph Beuys' (1974) definition of an association, konglomerat e.V. defines itself as a community of creative cultural producers, who work together on a "Social sculpture". The different workspaces are not separated or isolated from each other but merge with each other in different daily opportunities. The living culture of interdisciplinarity avoids and reduces hierarchies, alienation and separating work behavior. This includes positive side actions like collaboration and constantly developing products and processes. A space of mutual inspiration is created where openness is preferred, in contrast to patenting. The different workshops connect artists, autodidacts, young and old people, individuals and groups, professionals and hobbyists. Everybody holds responsibility and a bond of affection for their workshops. They organise themselves in an autonomous manner and set up their own rules depending on best knowledge and experience. The interface of the workspaces as well as the organisation is founded in the space itself.

The Story

The story of the konglomerat e.V. already begun long before the official opening in 2012. Since 2008 a couple of engaged individuals started to organise cultural events e.g. Räubertage, a music and do-it-yourself festival (www.raeubertage.) or Loophole Days of media (www.loop-hole.de). Based on these experiences the Werk.Stadt.Laden ("creation.city.shop") was founded in 2010. The Werk.Stadt.Laden was established to offer a free space to experiment and operate and included shared equipment as well as a serigraphy workshop. Several exhibitions, workshops and initiatives e.g. the RepairCafé, bookbinding or a diy laser developed in and around the Werk.Stadt.Laden. In 2012 finally the konglomerat e.V. was founded as an umbrella organisation of both the Werk.Stadt.Laden, FabLab and several workshops and independent projects. The goal was to stimulate and support own initiative and societal participation. To do something by yourself in a self-determined way and to work in cooperation was the principal idea.

In 2015 the konglomerat e.V. moved to its current location and the space is in a constantly developing process. Even now new workshops are created, e.g. the plastic recycling and the photo workshop.

Projects

Printing Monday: An Open Workshop as a Space of Encounter

Since 2015, our open silk screen printing workshop provides opportunities to meet and welcome refugees. This takes place under the Verbundstreffen offener Werkstätten ("networkmeeting of open workshops"). Starting with the network "Welcome in Dresden", the project is now located in the Dresdner Theatre. Practicing the concept of a mobile workshop, one Monday per month the organizers take paper, colors, screens and other utensils to the theatre. The "Printing Monday" is already quite a famous event which attracts not only refugees but interested locals as well. Under first instructions and some help, the visitors print their self-brought bags, t-shirts and pullovers by themselves. Frequently heard phrases are "You‘ve been here before! Could you explain this to him/her please?", "Could you please wash the screens? Ahh where are you actually from?". At these evenings bridges and binds are built between different persons.

For the organizers it is important not solely to offer a workshop where members can produce objects but also to practice something extra e.g. a warm welcome culture and a meeting space. At the end they want to show an alternative way of how to deal with an "avoidable" crisis.

Kunststoffschmiede: The Process of Plastic Recycling as a Manufactory!

At the beginning of 2017 a group of students contacted the space and told them about their idea to recycle plastic. This coincidentally confirmed the idea of the association to build up recycling machines as well as a project to include this process in the neighborhood. Therefore they built up the conditions to fulfil the first emerging needs. They were impressed with the ideas and a new workshop was quickly born. The basic idea of the plastic recycling workshop was to change the value of waste as an upgrading of the material and to increase the awareness of local processing. Therefore open workshops and talks were planned. Based on the idea of designer Dave Hakkens (www.davehakkens.nl), who provides the world with open source plans for recycling machines. The concept includes the whole process of getting the already washed waste/plastic from the neighbourhood and to shred and build up new objects in the workshops.

CHILLDUMP: Renewal of a Czech Skatepark With Young Local Residents

In cooperation with KulturOrt e.V. (http://www.kulturort-ev.de/) and the project "Kunst verbindet Nationen" ("Art connects nations") the space hosted a German-Czech exchange project. The main goal was to support teenagers in learning to help themselves. This was realised through an on-site workshop to build a new skateboard ramp while showing them the correct usage of the tools and the implementation of their own ideas. Prior discussions clarified how the skateboard ramp should be realized so that it would fit their vision and needs. Again, the project mirrors the concept of a mobile workshop where the association planned, packed and offered the tools and materials. Furthermore, the team conducted silk screen printing courses with Czech teenagers which in return taught the visitors of a subsequent festival how to use serigraphy. While teaching craftsmanship via the language of making, the exchange of ideas and thoughts between different cultures was achieved.

Velozack: DIY Cargo Bike Trailer

Velozack was designed by members of the Fablab Dresden in 2014 as a part of the metafab project (http://archiv.fablabdd.de/category/metafab/) connecting different spaces that utilize high tech, even in places as unapparent as an art gallery. The aim was to create a trailer to transport cargo easily and safely by bike, aspiring for stability, good driving qualities and sustainable materials. Thanks to its modular concept, the trailer is simple to build. as the frame is screwed together instead of welded. Just a chop saw and a electric drill are needed. The top plate can be adjusted and scaled to the users' needs (while the scale is mostly measured in beer crates). Besides this, the materials of the Velozack are affordable and use a standard design clutch from a commercial bike trailer company. Old bike parts like a pair of smaller wheels could be reused in order to save money and spare the environment.

Konglomerat e.V. is offering workshops where attendees can adapt, manufacture and assemble their own trailer within three days. This way people learn how to build their own products thus can repair or redesign it when needed. Through building things on your own you get a deeper understanding of how things (should) work and appreciate them more, hence handle them with care and saving resources.

Above all, the part list, CNC templates and construction plans are shared as open HW under a Creative Commons licence, encouraging people to build their own trailer, give feedback and enhance the design. Moreover finished trailers get shared within Konglomerat e.V. members.

RoseGuarden: OpenSource Entry System

Many makerspaces and open spaces struggle establishing an applicable entry system with regard to different aspects like safety and usability. There are always too little keys, their replication is costly and it's hard to keep track of them and manage the entry system in a way which is in accordance to the governance system of the association. Not to mention the expenses when a key gets lost. The project RoseGuarden depicts a social open source solution to these issues which can be adapted to specific needs.

The system is built upon a Raspberry Pi (https://www.raspberrypi.org/) that is connected to a RFID card reader, a relay and a power module, to switch the door closer. Users can be managed via a web user interface which is accessible within the local WLAN. It is possible to configure individual access times. Each user gets a RFID card for a fragment of the cost of a classic key. So new members just get a new RFID card associated to their access profile and thereafter can use the open space independently. Simply holding the card against the RFID reader, the Raspberry will check your access rights and the relay will switch the door closer. Talking about the costs, the expenses for establishing the system for one door is between 50-70€. One card costs 0,30€.

Therefore, the self-developed system fulfils exactly the needs of the space and can be easily reproduced to solve the entry problems at other places at relatively low cost in contrast to commercial systems.

RoseGuarden was presented in the Make magazine as a community project and after two years since its release, the project is already in use at seven other places throughout Germany.

The Importance for Multifactory Model

konglomerat e.V. was very important to explore the relationship between community and the technological devices used to manage it, and how to balance the need to automatize some processes and to avoid the feeling of centralized control, and konglomerat e.V. was fundamental to boost the development of some tools, like the e-window. konglomerat e.V. was also very important to see what happens to the community when the project scales up and the importance to balance skills and efforts when applying to a public call or a grant.

THE MUENCHEN FABLAB: MUNICH

This paragraph is based on personal communications that took place between August and September 2017 with Felix Tymcik, member of Muenchen FabLab.

The Munich FabLab started in September 2010 as a weekly meeting as guests in an urban cooperative workshop to do some tinkering – just three guys, a self made 3D printer and a lot of passion.

They managed to get a laser cutter funded, provided that it will stay in a permanent, lockable site.

The first place was a basement of about 45sqm. The few founders set up a program that would allow the biggest possible membership growth for a reasonable price, allowing engagement and propagation of knowledge to drive the initiative.

Munich is a rich, competition-driven and conservative town. The mainly hi-tech approach turned out to be very successful as opposed to the more alternative (partly much older) workshops.

In Summer 2017 the Munich FabLab counts about 350 members – the Munich FabLab hits the spot between a pay-to-go workshop and a community, with a core of people running the place and a much larger group of silent members who come on occasion and keep supporting the place with their membership.

When the number of members approached 100, the founders were already frantically looking for a new place in Munich, where space is rare. They moved to a quite centrally located commercial park, expanding to a whopping 400sqm. They now have a metal workshop (their dirty area) recently turned more into a wood workshop due to demand, an electronics corner, a large multi purpose area, a laser room where they hope to – in addition to a 30W Zing work horse – get a big 150W Monster to work on soon, one room dedicated mostly to 3D printing, vinyl cutting and screen printing, and not to forget – a chill out corner with a tall drinks-refrigerator.

After moving several times in recent years the founders kept their „old“ place as teaching area which is now separate from the workshop. Another very important trait of the Munich FabLab is the education program – they do single lectures and all year round courses for school classes, given by members happy to teach children something hands-on. This also is an income generator, as programs are partly funded by charities and companies.

Sometimes the place is social and very loud, sometimes weirdos occupy a lot of space, silently doing weird stuff or it can be even louder… so it is an ever changing, vibrant place.

As often the case in technical communities, it started almost as a bachelors club (one founder's wife joined in soon), but diversity is key for the success of a FabLab – now a third of the users are female:)

A big professional sewing machine caused quarrels with some boys favouring more technically advanced stuff, but it was democratically established as „required tool“.

They have a weekly members meeting and every two to three months big meetings for bigger decisions.

OZ OFFICINE ZERO: ROME

This section is based on personal communications that took place between September and October 2017 with Alessandro Sarte Splendori e Manfredi Scanagatta founders of OZ.

One of the great deceptions of our society is convincing us that simple things are actually complex. Contemporary societies, particularly Italian society, are characterized by a complex and technical bureaucratic system, which is often far removed from actually being useful for its citizens.

Bureaucracy is static, not dynamic. Social innovation, when it is expressed from the bottom, is revolutionary by nature, expressing new visions and actions that require a paradigm shift.

Winning is the innovation that uses existing rules to develop the process of change and once this change is effective, old norms generate new ones.

Institutions should play the role of innovation helpers by overcoming bureaucratic issues. Too often, top-down relationships structured between rulers and citizens limit the development of collaborative processes, therefore preventing the construction of participatory paths that involve its citizens as well as public and private entities.

To talk about social innovation from the bottom, it is necessary that strong relationships, normally imposed by political and economic forces, are counteracted by the feasibility and by the innovative value and social impact that the projects are able to develop.

A research and development model must be built, primarily based on the relationship between citizens, institutions and the private sector. Active citizenship must be able to be involved in decision-making processes through standard validation procedures, and not just through the call system.

Innovating society through new development systems of work and cities is a process that can not take shape if it is the offspring of a unique vision, whether this is dictated by a community, an economic or a political interest.

These are some of the issues they have been addressing for four years now in the construction and development of Oz-Officine Zero.

Oz-Officine Zero is an urban regeneration project and collaborative economy that has transformed the former RSI night train repair workshop into a multifactory, where today, 40 workers work within a collaborative system.

The project, run and supported by citizens, is developed in an occupied private site, whose destiny is linked to a bankruptcy auction, which will be soon announced by the court.

Both private institutions and communities are in a complex relationship that can either run out, develop through conflicts or through practices of collaboration.

Working on both the urban and working process is a choice dictated by a precise model of social innovation.

Every city has in its own spaces, big or small, that for several reasons have exhausted their original task. It is required to identify new possibilities of using urban sites, with a recycling vision rather than destroying and rebuilding, as this brings valuable ecological and economic benefits.

Limiting the production of complex disposal waste, cost-effective advantage is generated for those who will not have to bear the costs of waste management and an ecological benefit is also achieved for the entire community which will not have to bear the social costs that the waste generates.

What is even more important is the work environment: developing collaborative systems in shared spaces means that a freelancer is no longer alone, but integrated into a system that allows to exponentially implement the work of each worker through sharing skills and by building a common project, with respect to the individual needs.

CONCLUSION

As it is easy to see, the style and the communicative modes of the different spaces are very different and largely reflect the motivations, the history, the mission and the ambitions, the values of the single spaces, even before the possible characteristics of the models chosen to develop them. This, once again, should make us understand how varied and fluid the scene is, but also suggest how difficult it is for a group of initiators to decide which model to refer to and follow.

REFERENCES

Building Bloqs News, London Real Estate Forum - Message for The Mayor. (2017, June 19). [blog post]. http://www.buildingbloqs.com/blog/london-real-estate-forum-message-to-the-mayor

Chakrabortty, A. (2016, March 29). I've found the key to Britain's recovery: an orange shed in Shanktown. *The Guardian*. Retrieved from https://www.theguardian.com/commentisfree/2016/mar/29/key-britains-recovery-orange-shed- shanktown-social-enterprise-building-bloqs

Member stories. A long journey ends, another begins. (2017, June 20). [blog post]. Retrieved from http://www.buildingbloqs.com/blog/zana-agafonva-fashion-design-student

Senge, P. M. (1990). *The fifth discipline: the art and practice of the learning organization*. Penguin Random House Canada.

Skinner, B. F. (1948). *Walden Two*. New York: Macmillan Co.

Trautvetter, A. & Dalichow M., Junge A. (2012). Das kulturzentrum FreiLand in Potsdam Ein Experiment auf 10.000 quadratmentern [The cultural center Freiland: An experiment of 10,000 square meters]. *Projugend, 2012*(4), 25-28.

Chapter 6
New Needs Require New Spaces:
The Multifactory Model – Test, Phases, and Applications

ABSTRACT

The chapter focuses on the multifactory model, its evolution, and a step-by-step guide to building a multifactory. The first part of the chapter will present the reasons that led the authors to the decision to elaborate a model for a co-manufacturing space, the main features that characterize a multifactory, the key points of the multifactory model, and a description of the salient elements that have characterized Bigmagma, the development environment of the multifactory model. The second part presents the result of the evolutionary path of the model, presenting the subdivision into phases and macrophases of the intervention model and describing the first complete experience of application of the multifactory model, R84 Multifactory Mantua.

WHY A MODEL

Vision without systems thinking ends up painting lovely pictures of the future with no deep understanding of the forces that must be mastered to move from here to there. Peter Senge (2006)

DOI: 10.4018/978-1-5225-7089-9.ch006

During the first phase of the on-field research, the authors got in contact with shared workspaces that hadn't been developed following an existing model, a defined path. They were based on assumptions coming from some previous experience of the coordinators or initiators, and almost totally made on a trial and error basis. People belonging to different shared workspaces never met each other and didn't know anything about other similar projects. So, it was somehow surprising to discover that the shared workspaces faced the same problems and they presented many similarities in structure, organization and governance. Moreover, also the self-perception of people involved in different shared workspaces is very similar, as is the overall ethical point of view.

Focardi and Salati (2015) point out that "None of the spaces included in the research wrote anything about their experience, and they never theorized anything on their intervention scheme. Sometimes they have some theoretical refer- ence models coming from the US, but these models were completely changed as the socio-economical environment is radically different. As a matter of fact, all of the analyzed Multifactories are managing their growth by themselves, as if they have no idea that there are other similar places to compare with." And they also underline that "it was also clear that the lack of reference models leads the spaces to carry forward their own projects on a trial and error basis, so a systemization of these experiences and the development of a coherent model should have been very helpful to support the establishment of new ones, especially because places similar to the eight included in the research are in project or under construction, or open to the public every month, and a unifying intellectual framework could help their development."

Senge underlines that "personal vision, by itself, is not the key to releasing the energy of the creative process. The key is "creative tension," the tension between vision and reality. The most effective people are those who can "hold" their vision while remaining committed to seeing current reality clearly." All the spaces visited during the research tried to reach this balance between the visionary prospect of a future where human and professional relations are based on exchange and sharing and a present where economic operators can apply these principles concretely and at the same time be able to operate in highly competitive markets.

Sharing a vision and objectives is the basis, but it cannot be enough. Senge points out that "a shared vision is not an idea. It is not even an important idea such as freedom. It is, rather, a force in people's hearts, a force of impressive power. It may be inspired by an idea, but once it goes further—if it is compelling enough to acquire the support of more than one person—then it is no longer an abstraction. It is palpable. People begin to see it as if it exists. Few, if any, forces in human affairs are as powerful as shared vision. At its simplest level, a shared vision is the answer to the question, "What do we want to create?" Just as personal visions are pictures or images people carry in their heads and hearts, so too are shared visions pictures that people throughout an organization carry."

The keystone, therefore, lies in the organisational structure, which is the framework capable of making the shared vision of a group of people effective. The authors focused on those characteristics that could help to start, run and develop a comanufacturing space, run as a community project and highly heterogeneous in terms of members composition. The authors first defined a Model to describe the structure of a multifactory, taking from the observed shared workspaces the main elements in terms of governance, organization and composition, and then developed and tested the Intervention Model, a step by step guide to help in the construction of a multifactory from scratch and to support its development.

Shared workspaces are social systems, and as in any system the main point is not to start dynamics, but to manage them and this can only be done through a comprehensive vision of the system. This is the reason why it was necessary to develop a complete model. The model should be taken in its entirety, without attempting impromptu corrections, because many choices are difficult and counterintuitive, if evaluated locally, and go against common thinking, but are meaningful within an overall scheme.

The model was developed to help starting from scratch a multifactory ranging from class 1 to class3, following the classification provided in chapter 1, as these are the most common sizes within shared workspaces and can be started and run also without the need to involve huge companies or banks.

The intervention of originating a multifactory is structured in seven macro-phases. The macro-phases are numbered following an order that is both logical and corresponding to a time sequence.

In turn, macro-phases are divided into 30 phases. These phases are numbered following a logical order, but in practice the phases belonging to the same macro-phase can overlap over a temporal point of view.

The whole process includes a constant and continuous activity of supervision by the initiators of the project, involving both collective and individual meetings.

Therefore, the project is always under control, although it is fully participated and defined by the same companies as the Multifactory and is subject to constant revision and generation of scenarios, updated while the process develops.

MAIN CHARACTERISTICS OF A MULTIFACTORY

A Multifactory takes form as a shared workplace in which various types of companies coexist with each other.

Its members are very diverse and every company has the exclusive right to use part of the huge workspace and owns its means of production. These organizations are independent and work in full autonomy from each other, but are also structured in a way to facilitate close cooperation with each other and are able to operate as if they were part of a big company, becoming competitive at both the local and global level.

There are also many common spaces, the use of which is decided by the community by itself. Companies share skills, ideas, means of production and work together on common projects.

Each company pays a fixed monthly fee depending on how much private space is using. The fee is comprehensive of the running expenses as electricity and heating. Common facilities (like bathroom, kitchen, common desks space, show and event room) can be used for free and with the same rights by every company, independently by how many lots it occupies.

This simplified way of sharing costs not only makes it easier to keep the accounts, but also influences relations and fosters exchanges, as it makes it very clear that it's a place only for people who are willing to share, and helps to avoid conflicts when new companies want to join the project.

A Multifactory is a community project ruled by horizontal relationships, with no separation between "managers" and "users". Every company is part of an Association of companies, which is the managing entity of the space. In this way, every company, as a member, is responsible for the space in his ensemble.

The president and other members of the board have to officially represent the multifactory towards the outside, but decisions are taken by the community and all the managing tasks are distributed between all the members, and this is a solution to avoid that members should self perceive as "users".

Key Points of the Model

The Actors

Litefactories

The typical size of the enterprises part of a Multifactory ranges from micro to small. They are open to new opportunities and able to change very quickly, depending on the needs of the situation. They are flexible, innovative, sustainable and "thin" in respect of size, strategy, operational and structural aspects. Their effectiveness is mainly due to their intangible resources and also to a mix of good quality, innovation, customer care and a network of partner companies.

These companies by structure, style and perspective are positioned in a grey area between the world of the handcrafts, of the freelance and of the SMEs. It is a new class of economic actors that refer to productive categories coming from the past, but at the same time express specific traits. To identify these companies in a word, the choice was to introduce the term "Litefactory". It effectively expresses the "lightness" and flexibility of these economic operators, united in their inclination to act in the field of production.

Internal Stakeholders

Internal Stakeholders are entities directly involved in establishing, developing and managing a Multifactory. In 2015 Authors defined these internal stakeholders:

- **Initiator:** Who has to initiate the process that leads to the birth of a Multifactory, to write the project, to create consensus, to analyze the socio economical situation, to choose the place where the Multifactory will be established, to make the first selection of Litefactories.
- **Strategist:** Who has to define the overall strategic guidelines, objectives, and actions to achieve them, seek funding.
- **Formal Coordinator:** Who has to organize meetings and Plenary Sessions, to coordinate the working groups on joint projects, monitor the progress of joint projects, to solve conflicts.
- **Administrator:** Who has to pay bills, to choose, monitor, and pay services (surveillance, cleaning, waste charges).

- **Supervisor:** Who has to solve difficult situations of Litefactories, to collectively pool negative situations, to monitor the development phases of the Multifactory.
- **Institutions:** Public or private institutional stakeholders, as municipality, foundations, district governor, other public entities.
- **Owner:** Who has the right to properties on the area where the Multifactory is established.

Governance

It is essential that governance is participatory. This means that all rules, regulations and forms of control and coercion are decided by the members as a whole. Rules are established when there is a concrete case to discuss, and are the result of a confrontation between members and a mediation between different moral visions, which come together in a common ethic. This means that the set of rules necessarily has an incremental structure and changes with the change of members.

Structure

The structure must be far from the typical triangle organizational model, proposing on the contrary a horizontal solution with different areas of parallel responsibilities.

Managing Entity

The managing entity is the management structure of the multifactory, which is responsible for regulating and managing daily activities, defining internal rules and ensuring the maintenance and fair distribution of common resources among members. Main Tasks are listed in the Table 1.

The management structure of the multifactory may or may not be formalised, but the key point is that it must be organised in such a way as to ensure the full involvement of all members in space management activities. The recommended solution is an association of which all the companies members of the multifactory are also members. The association's statute must be structured in such a way as to give greater importance to the general meeting of members.

Table 1. Main tasks of the managing entity and involved activities

Tasks	Main Activities Involved
Ordinary administration	To pay bills, to pay services (surveillance, cleaning, waste charges).
Non-ordinary administration	To make the calls to fill the space, to sign the contracts with the Litefactories.
General services	To choose and monitor surveillance, cleaning service, etc…
High level coordination	To organize meetings and plenary sessions, to coordinate the working groups on joint projects, monitor the progress of joint projects, to solve conflicts.
Everyday coordination	To coordinate the use of common areas, of the kitchen (supplying, cooking), of shared tools.
Events organization	To decide, organize, and manage common events and exploitation activities.
Supervision	To solve difficult situations of Litefactories, to collectively pool negative situations.
Strategic planning	To define the strategic guidelines, objectives, and actions to achieve them, seek funding, to define new entries, to decide which joint projects to start, to decide marketing and promotion, to decide common actions.
Routine maintenance	To solve small practical problems (leaking pipe, lock to change)
Emergency maintenance	To solve big practical problems (broken heating, broken electric gate).
General rules setting	To set rules about space allocation, opening hours, allowed activities.
Everyday rules setting	To set rules about common area booking, kitchen usage, shared tools booking.

Coordination

The coordination of activities should be left to a third party in relation to the Management Entity and has the main task of organising activities and projects involving members. It can be a formal subject, such as a committee, or an informal subject, such as a working group.

Bigmagma Multifactory: The Testing Phase of the Model

While developing the Multifactory Model, the Authors put together the best practices they found during the first step of the Research, defining the basic idea of "Multifactory" which was then applied to the Bigmagma experiment. Bigmagma was the chance to set up many tools to manage the community and to help members to share and to collaborate in a structured way. The physical space occupied by Bigmagma was divided into two main areas. The Masterminds area was typically dedicated to desk activities, while the Manyhands area was dedicated to practical professions. It follows the presentation of the two areas that could be found on the site of Bigmagma:

Bigmagma as It Was Presented on Its Website

Following the same logic proposed above, Bigmagma's experience will be presented through the tools of self-representation and communication, in this case represented by the texts available on the website and the manifesto. As we can see, Bigmagma already had all the characteristics of a multifactory, starting from the heterogeneity and the mixture of professions, with a part of office work and a part of object production, to get to the exchange and mutual help, and a horizontal governance. Bigmagma also introduced some tools that were then inserted into the multifactory model and experimented with the structuring and implementation of the invisible factory.

Masterminds

the first area was called Masterminds and was mainly dedicated to the use of computersand the holding of meetings. was not to be understood as a coworking space, but as a common area available to members to carry out their office activities, whether they were design projects or administrative or communication. On the website it was presented as follows:

The main point of Bigmagma is Masterminds, the multifunctional space dedicated to those who use paper, pencils and computers. Masterminds changes shape according to your needs:

shared workspace, course space, meeting room, music space, project area. Masterminds is designed to work together, discuss, give birth and develop ideas. Masterminds is equipped with a large table made by TotoDesign, refrigerator, relaxation area, lockers, multifunction colour Wi-Fi laser printer with photocopier scan, B&W laser printer, white board, soundproofed cabin for audio-video editing or for reserved phone calls.

Masterminds generates stimuli, helps to produce thought and its view of the urban forest of the ex-Falck is a priceless source of inspiration.

Manyhands

The second main area was called Manyhands and was the area dedicated to the production of physical objects. The organization was based on a shared use of some spaces and equipment, with work areas of individual members for exclusive use. On the website it was presented as follows:

Bigmagma provides its members with Manyhands space, a fully equipped laboratory workshop. Manyhands can be used both for prototypes and small productions. Manyhands is equipped with a complete range of power tools and hand tools to meet (almost) every need: woodworking, metal processing, electronics, workshop cycle

Are you a designer and you have to make prototypes? Are you an architect and you have to build models? Do you take care of set-ups or sets? Are you a light craftsman looking for a place to make your own objects? Manyhands is designed for you.

The Invisible Factory: Large Projects

Bigmagma was also an opportunity to experiment with the activation and explicit management of projects organized according to the invisible factory mode. An example, that is the planning and launch of the "Grinding Project" on behalf of Nuova Ferrari & Zagni can be found in Focardi & Salati (2016). The concept of invisible factory was thus presented on the website:

Do you need to renew your customer base? Need to review your business or communication plan? Do you have internal communication problems in your organisation? Do you have to refit your premises? Do you have an innovative idea and want to develop it? Let's talk!

Bigmagma is a multifactory made up of very different professionals. This enables us to deal with wide-ranging projects.

Our areas of expertise are diverse: Collaborative economics, participatory processes, marketing and communication, HR, training, coaching, organisation and event management, Brand repositioning, product and process innovation, home design, industrial design, 2D and 3D drawing, prototyping, illustration, exhibition and shop set-ups. If you have a big project in mind, we will form a team of professionals able to follow the various stages of its development.

We are not a big company and therefore we don't have to bear the fixed costs of a large company, we are an invisible factory, we train teams and work on projects involving the most suitable and best skills for each individual project.

The Multidimensional System Design

To design, structure and organize projects coming from the Invisible Factory, Bigmagma developed a design methodology for projects, within the theoretical framework of the Community Economy, called Multidimensional System Design. On the website it was presented this way:

This design methodology has been developed with the aim of identifying a form that allows companies to address long-lasting projects minimizing risks, which must remain constant throughout the course of the project.

In particular, the basic concepts behind the methodology are:

- *Parallel Objective Design*
- *Incremental Design to Independent Blocks*
- *Constant Investment Design Development*

Parallel Objective Design

This design criterion implies that any project, of any nature, must pursue explicitly objectives that are far apart from each other.

The underlying idea is to reduce and dilute risk by finalizing a single investment in achieving different goals, which may also have different levels of priority.

Incremental Design to Independent Blocks

It means that a project must be phased in, independently, in which each one is autonomous and fully achieves its goals by allocating resources only for it. each step constitutes an addition to a block that, as a whole, becomes a predecessor.

In practice, future actions are based on the past, but the past actions impose what to do in the future.

This means that the project can be interrupted, modified or refurbished at any time, avoiding the need to re-work and achieving effective investment protection from the paradigm shift in the industry.

Constant Investment Design Development

This principle is based on the assumption that the project, once started, is part of the phase-by-stage investment, at least partially.

This means that the project does not provide for a large initial investment, from which time it goes back to, but an investment to which, as the project develops, is accompanied by a constant return from the very beginning, company exposure to a constant level, around 15-20% of total investment.

The goal is to build long-term projects, therefore they involve a high investment, but reduce the risk of enterprise to a minimum and avoid the need to resort to large, difficult-to-pay and expensive loans.

Of course, this requires a radically new approach to thinking about projects, taking advantage of the opportunities offered by more economical paradigms (sharing economy, network economy, crowd funding).

This is a versatile design methodology and can also be applied to promote and enable the development of individual companies in a Multifactory, helping them rethink their growth model and reduce business risk.

Through it, a Multifactory helps companies not only provide them with an optimum environment for their own development, but also providing effective methodological support for the launch of large-scale projects.

Collecto: A Practical Tool Developed by Bigmagma

Sharing Economy is a theoretical platform within is possible to imagine original models and methods for producing and working in a collaborative way. Comanufacturing Spaces, flexible and open, give the chance to test them.

One of the key themes that arise in these environments is the generation of the value of a service or of a commercial action in general. Where and when is value generated? How to pay for the different professionalism involved? How to weight value of the work of people who take part in a project?

These are wide-ranging questions that are worthy of generating extremely complex theoretical structures, but are also questions related to everyday life, specifically to professional daily life. Working in shared environments often leads people to face common projects, where they usually involve disparate professionalism, almost always related to distant professional areas.

Collecto is the system designed and applied by the multifactory Bigmagma of Milan for the distribution of profits from common projects.

Collecto applies when the project is already known in its general structure by members of the workgroup, sharing its overall objectives and structure.

So, there is already a general view of the project and the workgroup is made up of people who possess the skills needed to complete it.

In other words, it is the classic situation in which a professional (or craftsman or entrepreneur) has to deal with a job, or develop his own idea and cling to a group of other professionals (or artisans or entrepreneurs) around the design idea.

The group discusses the possible project to be implemented together, decides that the project is feasible and that the group can realize it.

It is time to assign the individual members of the group different tasks and decide how to share the earnings, or pay the benefits. The time of "who does what and how much is paid".

This is always a delicate moment. First of all, there are often overlapping skills in the workgroup, and several people may have the same role, secondly because it is very difficult to decide on each individual's pay in a peer group.

In the case of Bigmagma, none of the usual revenue sharing systems seemed to be good: the wages of individual project participants were not tied to the time spent on performing each task or the canonical social prestige associated with each activity.

Using Collecto, the first step is to divide the project into individual tasks that can be assigned as homogeneous tasks and to which a role is associated. This is done in collective form. As a first step, all project participants are asked to imagine a phase-out of the project itself. In practice, each participant divides the project into actions, in its own perspective.

In this way, each one offers the working group their own image of what the project is and how it is structured. All images are different, often very specific in relation to their professional scope.

Once collected, a meeting is called to define a common design structure. At the end of this phase, a common structure of the entire project will be defined, divided into precise tasks and shared by the entire working group. Everyone knows exactly how many parts the project is, what are the goals and the criticalities of each party as each party contributes to the development of the project and how each party is subdivided into tasks.

The design structure is reported on a spreadsheet, which is distributed to all members of the workgroup.

At this point, each participant is asked to assign a percentage value for each task. All scores are entered into a single spreadsheet and the average assigned to each task is calculated.

The result is the representation of how the working group as a whole reads the relative weights of the different tasks and each task is paid on the basis of the specific view of that particular working group compared to that particular project.

This result is again discussed by the team. The process is repeated until it is accepted by everyone and at that point the tasks are assigned.

What usually happens is, in practice, that they are paid more:

- Roles clearly fundamental to the specific project (linked to core skills that are at the heart of the project).
- Roles that require difficult skills to find in the circle of proximity.
- Roles related to unpleasant tasks.

Percentage values are inserted in an automated system that calculates the group's earnings and redistributes it to members according to the percentage value of the task performed by each one.

Collecto is interesting because it pursues different parallel goals. First of all, it offers all participants in the project a clear and understandable view of the structure of the project itself, the roles involved, the tasks needed to complete it.

It also allows to collectively discuss the whole structure of the project before startup and incorporate any variations and suggestions. Finally, it allows collectively to define the value that the group attributes to individual tasks and guarantees to all members of the workgroup both the ability to influence the value scale and the transparent view of the salaries of all members of the group.

In addition, in the case of projects cyclically repeated, for each edition different tasks can be attributed to different members of the group (if they have the skills) without having to discuss again the redistributive logic of the gains and giving to everybody the possibility to decide whether and what tasks to perform.

Bigmagma Manifesto

In order to better understand where members of Bigmagma Multifactory (2014) brought the theoretical reflection about their experience, it may be interesting to read their Manifesto, which gives a first hand idea of their self perception. The manifesto has been written by the steering committee and some members of the association in the way back from the first gathering of self organized cultural spaces in December 2013 at Manifatture KNOS, Lecce (Italy).

Bigmagma Manifesto (2014)

1.1 We believe in the value of difference, which is an asset, not a hindrance or a pretext for conflicts and divisions.
1.2 We know that it is not enough to put together different people to initiate positive dynamics.
1.3 We want to build the conditions for this to happen.
2.1 We believe in freedom.
2.2 We know that the freedom you claim has to be equal to that which you are willing to give.
2.3 We want to experiment practices that increase the freedom of the individual in a context of increased collective freedom.
3.1 We believe in the ability of humans to stay and work together without the need of a vertical structure of command.
3.2 We know that forms of regulation of relations are functional to stay and work together.
3.3 We want to make sure that these forms are horizontal and self-regulated.
4.1 We believe that the only good rules are these that do not need to be written.
4.2 We know that the rules often help to solve conflicts, but don't remove the causes.
4.3 We want to give us only rules in which we recognize, that serve to give a shape to the idea of shared values.
5.1 We believe that the work is an expression of one's being.
5.2 We know it is not easy, it is not obvious and for many people it is often necessary to compromise with their own desires.
5.3 We want to welcome, support and promote individuals, groups and organizations who agree with this idea.

6.1 We believe that every action of an individual is an act of personal responsibility.
6.2 We know that the need to work often leads people to act beyond the limits of their own consciences.
6.3 We want to make sure that everyone, regardless of their experience, position or education can have a choice.
7.1 We believe in the free circulation of knowledge.
7.2 We know that acquiring this knowledge takes time, effort and sacrifice, and that this knowledge has value.
7.3 We want this value to be repaid in the same coin, knowledge in exchange for knowledge, through a continuous professional exchange and an active and direct relationship between individuals.
8.1 We believe that those who propose an idea you should assume the responsibility of putting it into practice.
8.2 We know that it is difficult to possess all the necessary skills to develop their own idea.
8.3 We want to create the conditions so that those who have an idea can find around them the resources to put it into practice.
9.1 We believe in the personal and professional flexibility.
9.2 We know that flexibility and precariousness are close relatives.
9.3 We want to create conditions so that flexibility is organic to a personal project for professional development.

What Comes Out of the Bigmagma Experience

Bigmagma was a way both to experiment solutions and to prototype some tools that could be part of the multifactory model. Bigmagma gave the chance to develop tools like Collecto, to lower conflictuality between members. Another important aspect that was experimented with Bigmagma has been how a group of different economic agents can approach a single customer with an horizontal approach during the course of a long lasting project.

THE INTERVENTION MODEL: A STEP BY STEP GUIDE TO BUILD A MULTIFACTORY

In this chapter the way to build a multifactory from scratch will be presented. The intervention model provides step-by-step indications for building a multifactory. It has been developed to provide a practical and reliable guide to help anyone interested in developing a multifactory from scratch. This includes both subjects who own a space and want to create a multifactory, as well as an existing community of people interested in a shared workspace and looking for a physical place.

Macrophases

A multifactory can originate both from a space of from a community. In case the process starts from an empty space, it's needed to find the community, in case the process starts from a community, it's needed to find the proper space.

The first case typically takes place when there is an owner of an empty space, should it be public or private, who wants to transform it into a multifactory. The second one takes place when, within a community of professionals, artisans or self employed people, the idea of taking on a place together starts growing.

In the two cases the intervention phases differ at the beginning, and then merge after the establishment of the multifactory.

Case A: Starting from the space.
Macrophase scheme:
Macrophases
A: Preparatory Actions
B: Search of companies to become members and proposal to the territory
C: Official starting and formal acts
D: Space Restructuring, Team Building and First Dynamic Checks
E: Conventional definition of rules, internal rules, governance structure
F: First activities in the area
G: Multifactory Standard Arrangement

Phases

Division of Macro phases in phases

A 1: Site Analysis

In this phase initiators and the owner of the space have to decide how to divide existing spaces in order to fill in the member companies, defining the need of interventions to the structure and security plan.

In this phase it's also needed to perform a forecast analysis of overall yearly consumption, including heating, power needs.

A 2: Territorial Analysis

The second phase is an analysis of the territory, to define who the local actors are, that means groups and organizations that are active on the territory.

It's also needed to define which are the strongest local activities, and if local traditional industries do exist.

Then, it's important to define what is missing in terms of economic activities in that specific territory.

It's also fundamental to discover which are the places of aggregation where people who could constitute the target of the multifactory are used to meet, like bars, fairs, festivals, events.

A 3: Analysis of Local Stakeholders

The analysis of local stakeholders is meant to define which are the groups, entities, public and private bodies that could be interested or in any way affected by the multifactory. This includes municipality and other public bodies, Chamber of Commerce, schools, universities, labor unions, political parties.

In this phase it's also important to understand whether there are similar projects on the territory, which means projects intended to address the same target.

A 4: Definition of Stakeholder's Involvement

After the analysis of who the local stakeholders are, it's important to define with them what could be their involvement in the new multifactory. Also, if at this point formal agreements are not needed, it's better to define with local stakeholders whether they are willing to help the new multifactory and in which way.

If there are similar places, it's needed to understand if they could be competitors or collaborators of the multifactory.

A 5: Building Institutional Relations

After understanding which are the local stakeholders and how they could be involved, it's needed to build institutional relationships. First of all it's needed to meet with all stakeholders who are linked with local institutions and define who are the local politicians most interested in collaborating with the multifactory. It's fundamental to seek for support from local politicians and institutions.

It is not necessary, not needed and not advisable to ask the institutions to play an active role, or to support with grants or public funding, but it is very important that they are supportive and understand the public value of the multifactory.

Almost in every Country there are still no specific laws to regulate these spaces and, as they are experimental, it's important that local public bodies are aware of what is happening, what the multifactory will be and how it will be run.

A 6: Definition of Profiles of Companies That Could Be Part of the Multifactory

The next step is to understand which are the kind of companies that could come into space and see if and how it is possible to have heterogeneity in the multifactory. It is not possible to establish in advance the kind of companies that will be part of the multifactory, however it's important to be sure that in that given territory there is at least the possibility of having heterogeneity.

B 7: Definition of Open Call to Search Companies to Become Members of the Multifactory

In this phase the initiators have to define the structure of the Open Call and to write it. The Open Call will be the main tool to find new companies and it's important to express the requirements to be part of the multifactory.

B 8: Communication to the Local Community

After defining the Open Call, it's time to launch the call and to spread it within the territory.

Two main tasks will be participating to local events and the use of Social Networks.

B 9: Companies Selection

After companies answer the call, it's very important to make an accurate selection of the first "Pioneers" that will constitute the initial group of companies of the multifactory.

As they will play a crucial role, it's necessary to be selective, and the best way is to encourage a form of Autoselection. This can be done stressing some points in the Open Call and during presentations:

The first one is Internationalization and the need to have an international mindset.

The second one is the fact that a multifactory is a Community project and that the members will have to give a contribution.

The third one is the importance of Innovation and the need to be innovative companies to be part of the multifactory.

Even if during the course of the existence of the multifactory, people who do not have these characteristics will enter it, it is important to start with a more motivated group, as they will have to build the basis of the multifactory, the name, the logo, the rules, the Governance, the destination of common spaces and many other important aspects.

B 10:Looking for the Missing Companies

it is possible that the Open Call does not bring forward all the heterogeneity that is necessary, in which case it is necessary to actively search additional companies that may be interested to contribute to the definition of heterogeneity.

C 11: Space Allocation

While the Open Call is active, it's important to offer guided visits to the spaces that will host the multifactory, to allow the group of companies to become familiar with space.

When the Pioneers have been defined, it's possible to allocate spaces according to different needs.

C 12: Work Plan Definition

At the initial stage of the multifactory there are much fewer companies than those that could be theoretically hosted. This can lead to some dispersion. One method to mitigate this dispersion is to keep the initial companies physically close one to each other, but it is always important to keep in mind the general view, so if there are companies that for various reasons have to be kept apart from others, it's a better to do it from the beginning. For example, companies which work with children, or which make noise, or have special needs.

C 13: Group Constitution

At the beginning, it's important to organize many team building activities by doing concrete activities such as organizing a cleaning day of common spaces.

C 14: Managing Body Foundation and Contract Signing

One of the first common activities is to define together which type of formal structure will be responsible for the space, should it be an association, a company, a cooperative, or any other kind of private body. The companies members of the multifactory will build the statute together and will define the Governance. It's important to notice that this phase comes after the Open Call and even after the allocation of the spaces.

D 15: Occupation of the Spaces and Beginning of Production Activities

Once the preparatory stages are over, companies can take possession of the spaces and start living the multifactory in the daily life.

D 16: Team Building

To create a strong group, it is important to organize a series of practical courses and meetings to define the communication main points. This includes the construction of the visual identity, and the choice of the name and logo of the managing body.

D 17: Collective Renovation

At the beginning it very difficult to fix common spaces because they are a lot and fixing them requires a lot of energy but also because the general style of the multifactory needs time to be defined. If someone within the multifactory deals with architecture, design or graphics they can use the opportunity to renew common spaces as part of their portfolio, and to experience new solutions. So, even before the multifactory has started, it's a chance for professional growth.

It is important to give the opportunity to those who have ideas of using common spaces to carry out projects independently, so that subgroups are created with a group leader who receives from the community the mandate to deal with developing a solution for certain areas of the multifactory.

D 18: Processed Attendance Review

The initiators should try to do as little as possible on common space and common events, trying to give freedom to the pioneers to express themselves and to shape the common spaces in the way that best represents them. At the same time, it is necessary to give the group the idea that the process is supervised so it is necessary to do several supervisory meetings, as the first phases of the process are very critical and it's absolutely necessary to be very careful and plan to provide professional support to the group.

D 19: Check Companies Suitability and Replacements

After the companies entered the spaces and started to work together, it's needed to make a series of meetings with individual companies to figure out whether they are actually able to start and be functional to the group or not. If not, it is best to leave the project as soon as possible to prevent them from losing time or harming the constitution of the group.

E 20:Identifying Initial Needs of Internal Rules

Rules are not decided before starting daily activities but are generated little at a time as a response to concrete cases. So after the start of activity it is extremely important to take care of marking all the events that require regulation. For example, how to handle the kitchen or common kitchen or the relationship with the internships.

E 21: Formalization of Governance, Standards and Regulation Structure

Starting from the solution to concrete cases, rules are formalized, and become a body of common norms that is the set of rules that everyone must adhere to.

E 22: Early Trading and Promotional Actions

It is important that member companies of the multifactory have a direct income and perceive the multifactory as a structure that can improve business performance.

So it's important to start a series of activities right from the beginning to support sales or actively support the growth of multifactory companies.

F 23: Initial External Communication Plan Definition

It is important to define an integrated external communication plan and make everyone participate, for example through the common management of the social network of the association.

F 24: Opening

The inauguration is a crucial moment and it is important that it is postponed until the companies have something to show and the space is beautiful.

G 25: Maintaining Institutional Relations

It is of fundamental importance to keep in constant contact with institutional stakeholders, to involve them in the projects and make them active part of the progress of the multifactory, and invite them to collaborate in joint activities, such as organizing events, engaging in invitations, seizing opportunities to invite them to the multifactory and make them meet with companies.

At the same time it's necessary to meet other shared spaces or collaborators on the territory (coworking, fablab, makerspace) to imagine possible exchanges and / or a real value chain

G 26: Long-Term Commercial and Promotional Actions

In the meantime, it is important to establish long-term commercial and promotional strategies that can help companies in a structural way, like agreements with other entities and foreign trade bureaus, or collaborations with other institutions.

G 27: Definition of Ordinary External Communication Plan

In meanwhile it's also necessary to define a Social Network activity and to Plan joint participation in fairs and events.

G 28: Structuring Relations With the Territory

Program and realize a series of activities, workshops and events with other people active in the area. Have good-neighbourly relations with local stakeholders.

G 29: Definition of Cultural Programming

The multifactory must become a landmark, and therefore must propose to the territory its own values, which are the values expressed by members companies, and therefore it's needed to do a number of activities to disseminate the corporate culture as expressed by the companies that are part of it.

G 30: Management and Procedural Autonomy of the Group

Time after time, the role of the initiators must always leave more room to the autonomous organization of companies that, once learned the operating mechanisms, can take advantage of the opportunities provided by the multifactory.

Time Schedule

The setting up of a Multifactory is quite fast, about nine months, a value that can change according to the specific requirements of the site, territory and companies. Any critical events occurring in one or more of the phases may lengthen the time of one to two months, even though it is unlikely that there will be critical events in all macro-phases.

The length of the macro phases can vary according to unscheduled events, but there are some main reasons that can increase the length of each macro phase:

A to 1 to 3 Complexity of site and condition of the site (age, storage, health and safety anomalies). Any special features of the territory and / or Stakeholders.

B from 1 to 3 Level of anthropization and entrepreneurial solidity of the territory. Presence of already active communication channels.
C 1 to 2 time availability for common meetings (especially for companies already started)
D from 2 to 6 Experience and resources of the Lite Factory involved.

And from 1 to 2 Previous collaborative experiences of the Lite Factory. Level of homogeneity of views, values and needs.

Economic Plan

The cost to start a multifactory on an area of about 1000 sqm is 120,000 Euros overall, provided that the site is well maintained.

This figure covers the entire development of the intervention, including any necessary redesigning actions and is fronted in the first two years by the owner of the site, and then repaid over a five year period by the companies part of the Multifactory.

This figure includes all the activities required for Multifactory structuring, Site Analysis, Call Layout, Real Estate Selection, Training, Supervision, Process Review, Overall Strategic Design, Marketing and Communication, Leadership of the project in all its phases up to the autonomy of the Multifactory and only the live expenses (such as leaflets, FB campaigns, graphics, catering, etc.) are excluded.

The project envisages that the entities included in the Multifactory at the beginning of their stay pay a fixed monthly fee to cover normal costs (heating, surveillance, etc.) and the cost of setting up the Multifactory itself. This quota is designed to contain the maximum costs for existing entities in the early years.

The standard economic plan envisages a three-year Free Rent, followed by an increasing rental fee.

The Free Rent is justified by the fact that in the initial stages of a Multifactory's life the companies that create it must invest as much as they can to promote the business as well as bear some costs for the renovation of the property. A Free Rent allows them not to be initially burdened by an unsustainable load and, ultimately, it is functional to the launch and territorial rooting of the Multifactory itself.

Being in a Multifactory allows them to grow and consolidate, but in the early stages it is necessary that tthe companies can invest their earnings primarily in marketing or business operations. This allows them to stabilize more quickly and, ultimately, is a way to stabilize the chances of generating space income.

Considering an average space, the annual fee for the first year would be around 3000 euros, including all expenses. This sum will rise to 3500 euro / year in the second, third, fourth and fifth year, a figure that also includes rent, since the entities begin paying it. At full capacity, or from the eighth year, it is expected that 75 square meters will cost, all inclusive, approximately 6000 euros per year.

This plan facilitates Start-Up activities while at the same time eliminating all fixed costs related to the facility from the outset and covering all of the startup costs of Multifactory in a short time, leading to make the structure generating income within three years from the start of the operation.

For the property, once underway this means a total annual rent of around 56,000 euros

For small companies, it means a monthly outlay, including any ordinary expenses, utilities and rent, which ranges from around 260 euros / month in the first year and reaches approximately 480 euros / month once at full capacity.

So the entire operation is largely sustainable for the property, of course in a non-speculative outlook, but also for the members, wjo can embed their business into the Multifactory with an irresistible initial outlay, which remains modest even once at full capacity.

This simulation was realized by assuming a scenario that favors cost containment. This choice, very prudent, has become the most appropriate considering the current weakness of the companies. Even in this context, the choice to start a Multifactory seems to be of utmost interest to the property. Obviously it is possible to hypothesize different scenarios, depending on the strategic options defined by the property for the site concerned.

THE FIRST APPLICATION OF THE MULTIFACTORY MODEL: R84 MANTOVA MULTIFACTORY

R84 is the first Multifactory developed from scratch following the Multifactory Model. R84 takes place in Mantova, a small capital of province located in North Italy, between Milan and Venice. Mantova used to be an industrial city, as after WWII several industrial plants were built close to the city following plans for post war reconstruction of the heavy industry asset of the Country.

Table 2. Example of multifactory sustainability plan simulation for 12 years

	Year 1	Year 2	Year 3	Year 4	Year 5	Year 6	Year 7	Year 8	Year 9	Year 10	Year 11	Year 12
rent	0	0	0	0	11250	33750	45000	56250	56250	56250	56250	56250
restructuring	0	0	0	0	0	0	0	0	0	0	0	0
managing	0	0	0									
heating	11200	11200	11200	11200	11200	11200	11200	11200	11200	11200	11200	11200
power	10000	10000	10000	10000	10000	10000	10000	10000	10000	10000	10000	10000
General expenses	5000	5000	5000	5000	5000	5000	5000	5000	5000	5000	5000	5000
supervision	0	0	0	1500	1500	3000	3000	3000	3000	3000	3000	3000
consultancy	25000	25000	25000	25000	20000	0	0	0	0	0	0	0
Yearly total	**51200**	**51200**	**51200**	**52700**	**58950**	**62950**	**74200**	**85450**	**85450**	**85450**	**85450**	**85450**
Of which excluded charges	*51200*	*51200*	*51200*	*52700*	*47700*	*29200*	*29200*	*29200*	*29200*	*29200*	*29200*	*29200*

Table 3. Example of cost simulation for single small slot

companies	15											
space mq	75											
total square meters given	1125											
Yearly rent for mq	0	0	0	0	10	30	40	50	50	50	50	50
Annual fee for 75 mq	0	0	0	0	750	2250	3000	3750	3750	3750	3750	3750
Contribution to expenses oer year	3413	3413	3413	3513	3180	1947	1947	1947	1947	1947	1947	1947
Yearly contribution plus fee	3413	3413	3413	3513	3930	4197	4947	5697	5697	5697	5697	5697
Total from fee	0	0	0	0	11250	33750	45000	56250	56250	56250	56250	56250
Monthly total for 75 mq	**284**	**284**	**284**	**293**	**328**	**350**	**412**	**475**	**475**	**475**	**475**	**475**

A huge refinery, the first one built in the river Po valley, started production in 1953. In December 2013 the last owner, MOL Group, announced the shut down of the refinery. The production lines were stopped in January 2014 and more than three hundred employees were fired. To manage the social crisis that followed, local government, trade unions and the property of the area started a process to foster the reindustrialization of the area, to attract private investors, and to implement innovative productive concepts in the site. Delegates from local government, Lombardy region, Italian government, trade unions, the environmental protection agency, and the owner of the area formed a commission to evaluate proposals to redevelop the area and launched a call for projects. The selection process of the economic activities to host in the area took two years and in 2016 two projects were allowed to start their activities. R84 Multifactory Mantova is one of these two projects and is a shared workspace shaped on the Multifactory Model. R84 Multifactory Mantova consists of six buildings (four two-storey office blocks, a workshop and a small warehouse), and covers an area of approximately 8,500 sqm including garden and parking lots. R84 Multifactory Mantova didn't receive any private or public funding and is completely self-sufficient, but the area is given on free rent for twelve years and the owner covers all costs for heating and electricity for first two years.

R84 Multifactory Mantova is run by an Association of Entrepreneurs, called "Associazione R84", which is made up by the same members of R84 Multifactory. The association is the managing entity of R84 Multifactory Mantova and is responsible for any financial, administrative and strategic aspect of the multifactory.

Every company that is part of R84 Multifactory Mantova pays a fixed monthly fee, based on the size of their office or workshop, which includes all expenses and services. An all-inclusive "flat rate" helps professionals and startups to stay focused on their business, as they can avoid unexpected expenses and save money over leased space alternatives.

R84 Multifactory Mantova has international relationships with other Shared Workspaces in UK, Portugal, Spain, USA, Ecuador, and Germany.

R84 Multifactory Mantova aims to promote and support activities related to crafting, the Makers movement, entrepreneurship, and in general art and professionals. Special attention goes to development of innovative products and services, and to investigate how to foster innovation through collaboration and a collective rethinking of creative processes.

In 2017 R84 Multifactory Mantova was partner of a huge project related to Augmented Reality and exploring new potentials coming from innovative fields, as robotics, smart manufacturing, industry 4.0, artificial intelligence, virtual reality and augmented reality.

How R84 Was Developed: The Macrophases

R84 Multifactory Mantova was developed following the macrophases foreseen by the intervention model described in the previous chapter of this book. The following is a brief description of each macrophase and the period in which it was carried out.

A – **Preparatory Actions:** This macrophase took action between February 2015 and April 2016, when the final agreement was signed.

B – **Search of Companies to Become Members and Proposal to the Territory:** This macrophase took place from May 2016 and July 2016. The deadline for the Open Call was July, 18th. At the end of the selection process nine companies were selected to constitute the initial group. Eight of them were selected by the Open Call (phase B9). They were a photographer, a consultancy company, a design studio, a theatre company, a painter, an association of consumers, an artisan and a stylist. An IT company was added to the group after a direct contact (phase B10).

C – **Official Start and Formal Acts:** This macrophase should have lasted up to two months, as the property should have delivered the area by September 2016, but the delivery was postponed until February 2017 due to a delay in the installation of the heating system and the removal of some asbestos pipes. This five month delay had a very strong impact on the initial group of companies, and five of them (the painter, the association of consumers, the artisan, the stylist, and the IT company) found a different accommodation. Two other professionals joined the initial group, and R84 Multifactory Mantova started its activities with six companies.

D – **Space Restructuring, Team Building and First Dynamic Checks:** This macrophase took five months, from March to July 2017. In the end, a company made the decision to leave the multifactory (the consulting company). Two more companies joined the group (an architect and a social cooperative).

E – **Conventional Definition of Rules, Internal Rules, Governance Structure:** This macrophase took two months, from August to September 2017.

F – First Activities in the Area: This macrophase took four months, from October 2017 to January 2018. At the end of this period two architects left the multifactory.

G – Multifactory Standard Arrangement: This macrophase had already started in February 2018, and by march 2018 eight more became part of the multifactory (another photographer, three artisans, a psychologist, a masseur, a naturopathist, a designer, a real estate company).

Sustainability Plan of R84 Multifactory Mantova

The sustainability plan for the R84 Multifactory Mantova faithfully reflects the general business plan of the intervention model. It has been developed over a period of twelve years, with the project being completely self-sufficient from an economic point of view from the second year onwards.

Governance System of R84

R84 takes form as an Association of Entrepreneurs. As an association, the Art. 8. Defines its social bodies. They are organs of the Association:

Table 4.

Worst Case	Y 1	Y 2	Y 3	Y 4	Y 5	Y 6	Y 7	Y 8	Y 9	Y 10	Y 11	Y 12
SMEs	6	12	21	24	30	35	35	35	35	35	35	35
% filled	10	20	35	40	50	60	60	60	60	60	60	60
revenues K€	12	24	42	48	60	73	73	73	73	73	73	73
adm. costs K€	2	0.5	0.5	0.5	0.5	0.5	0.5	0.5	0.5	0.5	0.5	0.5
maintenance K€	2	2	2	0.4	0.5	0.6	0.6	0.6	0.6	0.6	0.6	0.6
marketing K€	0.6	0.8	0.4	0.4	0.5	0.6	0.6	0.6	0.6	0.6	0.6	0.6
events K€	0.2	1	0.4	0.4	0.5	0.6	0.6	0.6	0.6	0.6	0.6	0.6
heating K€	0	0	4	4	4.6	5.6	5.6	5.6	5.6	5.6	5.6	5.6
electricity K€	0	0	1.5	1.5	1.6	1.9	1.9	1.9	1.9	1.9	1.9	1.9
cleaning K€	0.5	0.5	1	1	1.4	1.7	1.9	1.9	1.9	1.9	1.9	1.9
internet K€	0.5	0.5	0.8	1	1.2	1.5	1.5	1.5	1.5	1.5	1.5	1.5
running costs K€	5.8	5.3	10.6	9.2	10.8	13	13.2	13.2	13.2	13.2	13.2	13.2
Supervisors K€	4	13	22	22	26	32	32	32	32	32	32	32
fund K€	2.2	5.7	9.4	16.8	23.2	28	27.8	27.8	27.8	27.8	27.8	27.8
total costs K€	9.8	18.3	32.6	31.2	36.8	45	45.2	45.2	45.2	45.2	45.2	45.2

- The Shareholders' Meeting;
- President;
- The Board of Directors;

To guarantee the democratic nature of the Association's structure, it is established that all posts must be elective and, as stated in Art. 13, all social offices have a three-year term. Replacements and co-options made during the three-year period will expire at the end of the three-year period.

The Art. 9 defines functions and limits of The Shareholders' Meeting

The Shareholders' Meeting is the sovereign body of the association and is composed of all members in good standing, with the payment of the social portion and the annual contributions and which, at the date of the notice of convocation, are registered in the Members' Book.

The Shareholders' Meeting is convened by the Chairman at least once a year, and whenever the Board of Directors considers it appropriate and is presided by the Chairman or, in the event of his impossibility, by a Board member. However, the convening of the Shareholders' Meeting may be required by at least two thirds of the members, provided that the membership fee is paid in full.

The convocation of the Shareholders' Meeting shall be made by notice at the Company's registered office at least fifteen days before the date set for the first convocation meeting and shall contain the agenda, date, time and place of meeting. In the same notice of convening the meeting, a further day may be set for the second call.

The call is also made by e-mail, fax, letter confirmed by the recipient even with the same means. The associates, for the purpose of their relationship with the association, elect domicile at the place and at the e-mail address indicated in the Members' Book.

Video conferencing, video chat, social networking, telephone or any other system or device that allows members to participate actively and democratically in the Assembly is allowed.

The Ordinary Shareholders' Meeting resolves:

- The election of the Chairman and the Board, setting the number of members;
- The election of the substitutes of the members of the Board of Directors possibly leaving office;
- The approval of the financial statements and the annual report;
- On the issues raised by the Board of Directors;

At the first convocation, the ordinary assembly is regularly constituted by the presence of half plus one of the members entitled to attend; in second call it is valid whatever the number of members involved. The Ordinary General Meeting deliberates on the items on the agenda with the consent method as specified in Article 10. By way of derogation from this general rule, the Chairman may request, on specific decisions, to hold a vote . In that case, the Ordinary Assembly decides validly, both before and after the second convocation, by an absolute majority, ie by a favorable vote of half plus one of the voters.

The Extraordinary Shareholders' Meeting resolves:

- Requests to amend the Articles of Association;
- On the dissolution of the Association;
- On the appointment of the liquidator;
- On any other topic of extraordinary character and of public interest placed on the agenda.

Regarding amendments to the Articles of Association, the Extraordinary Shareholders' Meeting is regularly constituted by the presence of at least one half plus one of the associates, and resolves by an absolute majority. With regard to the dissolution of the association and the appointment of the liquidator, the Extraordinary Shareholders' Meeting resolves with the favorable vote of at least three quarters of the associates.

All minutes, shareholders' resolutions and statements, as well as being duly transcribed in the minutes of Shareholders' Meetings, are posted to shareholders with the exposure for 30 days after approval at the Association's headquarters and / or by sending to mailing list of the association.

To ensure the democracy and the complete involvement of all members in every decision, the Art. 9 states that every decision should be taken with the Consensus Method, as expressed in Art. 10

The principle behind the functioning of the Ordinary Assembly is the application of the consensus method. This method pursues a full and complete inclusion in the deliberations of the decision-making bodies of the Association of Viewpoints of each single partner.

The method of consensus states that each person's point of view must be heard on any matter and that a comparison is initiated so that all participants can be convinced that their views and concerns have been accepted in the assembly in the form of change of decision or insertion of explicit control or control systems.

There is no need for the unanimity of the favorable opinions on each issue, but the inclusion in the final decisions of dissenting points, even with monitoring function.

The decision is considered taken if a formal vote is needed when the chairman has exhausted the discussion on the specific question, he notes that there are no objectionable objections by any participant, he declares it to the Assembly and no explicit dissent. The method can be defined in its formal steps within the Association's regulation. As specified in Article 9, by way of derogation from this general rule, the President may request, on specific decisions, to hold a vote. In that case, the Ordinary Assembly deliberates validly, both before and after the second convocation, by an absolute majority, that is to say, with a favorable vote of half plus one of the voters.

To develop strategies decided by the Assemby, the Art. 11 defines the role of the Board of Directors. The Board of Directors is the executive and management body of the Association and is elected by the Assembly every 3 years. It consists of a minimum of 3 to a maximum of 11 members, including the President who is a member of the law. Board members are re-elected and all assignments are free of charge. The Board of Directors may be revoked by the Shareholders' Meeting; it will remain in effect until the election of the new one. In the event of the resignation of a member of the Board of Directors, the first non-elected member is co-opted. One or more Deputy Presidents, a Secretary and a Treasurer may be appointed within the Board of Directors. The Chairman, who has the legal representation of the Association, may be delegated some of the powers attributed to the Board of Directors. The Board of Directors has the widest powers for the ordinary and extraordinary management of the Association. The Board of Directors is in particular responsible for:

- The resolution on the admission of new members.
- The determination of the social quotas.
- The decisions concerning ordinary and extraordinary expenses, exercise and capital, for the management of the association; decisions related to the institutional, complementary and commercial activities and services to be undertaken in order to achieve the best institutional goals of the association.
- Decisions concerning the direction of the employees and the coordination of the collaborators and professionals employed by the association.

- The annual report of the financial statements to be submitted to the shareholders' meeting within four months of the end of the financial year; an external professional chosen by the board of directors • the preparation of the annual report on the activities carried out and the objectives to be submitted to the assembly.
- The presentation of a plan for the activities to be carried out in the new social year.
- The right to appoint, the external members of the council, the delegates for the performance of particular functions that are established from time to time by the board of directors itself.
- The drafting and approval of the administrative regulations and the proposals for amending the articles of association to be submitted to the subsequent approval of the shareholders' meeting.
- Each function that the statute or the laws do not attribute to other organs.

The Board meets at least twice a year, or whenever the Chairman or a majority of the members consider it necessary. Council meetings should be made with a written notice to be delivered at least 15 days before the date of the meeting; such notice must contain the agenda, date, time and place of the meeting. The meetings of the Board of Directors are in the sole convocation, valid with the presence of at least a majority of its members and are chaired by the Chairman or in his absence, by a counselor appointed by the present. In the event of the same number of votes, the Chairman's vote prevails. The sessions and the deliberations of the Board are made by a minutes signed by the Chairman and the Secretary. The Board of Directors declines for half-time resignation plus one of its members. In this case, the Chairman or, in the event of his being prevented, by the Vice-Chairperson or, in the alternative, by the Senior Adviser, shall convene the Extraordinary Shareholders' Meeting within 15 days and hold within the next thirty by the ordinary administration.

The role of the President is defined by Art 12. The President has the legal representation of the Association and can sign documents and contracts. The President is elected by the Shareholders' Meeting, together with the members of the Board of Directors, every three years. The President leads the Shareholders' Meeting and the Board of Directors and arranges for it to be convened, supervised on the performance of resolutions of the Shareholders' Meeting and the Board of Directors. In cases of urgency may exercise the powers of the Board of Directors unless it is ratified by the latter at the first useful meeting. The President may delegate, in writing and in writing only to one or more members, for both individual acts and categories of acts. In case of absence, impediment or resignation the functions of the president are carried out by the vice president of the association.

CONCLUSION

The on-field research has shown how the spaces of comanufacturing have grown following autonomous paths of development that are largely overlapping. The multifactory model was developed to systematize the key points of these paths and integrate them into a coherent development system. This development system is the intervention model, which explains the phases through which a multifactory is built from scratch.

An indication that has come ubiquitously from the experiences visited during the research is the need to define formal environments to encourage structured collaboration between companies, especially with regard to the division of tasks within a project, the value of different roles and the corresponding distribution of profits. Another is to define with particular care the governance structure of a multifactory, so as to safeguard the democratic nature of decisions and encourage the active participation of all members. In addition to what is reported in the model, the Authors wanted to add some specific indications regarding governance tools and that promote collaboration, which were taken from the concrete cases of Bigmagma Multifactory and R84 Multifactory.

REFERENCES

BIGMAGMA Manifesto. (2014). Retrieved from http://bigmagma.it

DESTROY POWER POINT! Primo incontro nazionale tra spazi per i giovani, la creatività e l'innovazione sociale. (2013, December 13). Retrieved from http://www.manifattureknos.org/knos/calendario.php?eventID=281

Focardi, G., & Salati, L. (2013, September). *Modeling the Future of Job Creation: the MultiFactory Model*. Paper presented at the ECCS'13 (European Conference On Complex Systems) sat. CODYM, Barcelona, Spain.

Focardi, G., & Salati, L. (2013, September). *MultiFactory Model: a New Approach to Job Creation*. Paper presented at the WES 2013 (Work, Employment and Society Conference), Warwick, UK.

Focardi, G., & Salati, L. (2016). Social Media as Elements of Shared Workspaces: The Multifactory Case Study. In A. Goel (Ed.), *Product Innovation through Knowledge Management and Social Media Strategies.* Hershey, PA: IGI Global

Servizi ed uno spazio aperto per. (2017). Retrieved from http://www.r84.it/

Chapter 7
Future Research Direction

ABSTRACT

This chapter gives a general overview of the main issues relating to co-manufacturing spaces and which could be the direction for further research. Crowdworkers are shaping a new model of society, fair, supportive, inclusive and based on participative tools. Co-manufacturing spaces are the places to realize and experiment this model, make it grow, and spread it. But there are still many weaknesses, risks, and problems to be solved, starting from education. In this chapter, these themes are briefly presented, as they are open topics and can be traces for future research.

A GLOBAL VISION: THE INTERNATIONAL NETWORK OF COLLABORATIVE SPACES

Peter Senge (2006) underlined how "Individual learning, at some level, is irrelevant for organizational learning. Individuals learn all the time and yet there is no organizational learning. But if teams learn, they become a microcosm for learning throughout the organization. Insights gained are put into action. Skills developed can propagate to other individuals and to other teams (although there is no guarantee that they will propagate). The team's accomplishments can set the tone and establish a standard for learning together for the larger organization." So, innovation is fostered by a collective propagation of skills, that become a shared value for the group. That said, it is also true that this transfer takes place through direct contact between people and Senge (2006) himself recalls that "It is important to recognize that mastering any of the

DOI: 10.4018/978-1-5225-7089-9.ch007

disciplines requires effort on both the levels of understanding the principles and following the practices. It is tempting to think that just because one understands certain principles one has "learned" about the discipline. This is the familiar trap of confusing intellectual understanding with learning. Learning always involves new understandings and new behaviors, "thinking" and "doing." This is the reason for distinguishing principles from practices. Both are vital."

In this view, the concept of "learning" is related both to the acquisition of new intellectual tools and to the ability to put them into practice. It is not only important to have theoretical knowledge, but also to know how to apply knowledge, and this is mainly through the repeated application of knowledge within a group. This is what Authors found to happen in shared workspaces, and especially in Comanufacturing spaces as there is the presence of manual workers who need to gain practical capabilities.

Comanufacturing spaces are based on the idea of creating connections between members and foster innovation and business ideas through cross-contamination between members, but not all the skills are available at a local level and the opportunities to Exchange at some point come to an end. During the European research an idea took shape that a network between different Comanufacturing Spaces could expand the opportunities for collaboration to a wider scale. While running Bigmagma Multifactory the authors established an International Network to connect Multifactories, FabLabs, Makerspaces and Open Workshops so that companies, artisans, freelancers and independent workers that are a part of them could cooperate in a stable, viable, and structured way. The network started in 2014 as a European experience under the name of "Multifactory Network", but since 2016 the network was also extended to the USA and Ecuador and is supported by the *Verbund Offene Werkstätten*, based in Berlin (DE).

The idea behind the International Network of Collaborative Spaces INCS is to connect some Comanufacturing Spaces all around the world in a kind of "brotherhood". Companies, artisans, freelancers and independent workers who are part of them can cooperate in a stable, viable, and structured way based on the idea of sharing and cooperating and can thus compete with capitalistic traditional companies.

Goods should be developed by international teams of independent workers, working from different spaces and connected with a dedicated communication system. Then, the goods could be produced by a shared network of small manufacturers, grouped in clusters of local or regional value chains. It is a beautiful concept which would need local agents to bet set up, and a fair amount of time. The challenge is to build a system that can work without a central structure and that, in a self-regulated way, can prove to be effective for the growth of its members.

A strong and structured Network would be a way to accelerate partnerships between equals and the ability to create strong and large enough teams to reach an international market, anticipating traditional companies and avoiding the replication of the capitalistic model when it comes to producing complex products.

Shared workspaces (Fab Labs, Makerspaces, Co-manufacturing spaces, and Multifactories) are changing the way goods are invented, developed and produced.

The Sharing Economy can give a formal framework to develop new concepts, but in everyday work life people need to be in contact, to share ideas and knowledge, and most of all to meet in person.

The Network and the Shared Spaces

Connecting companies which already operate inside Comanufacturing Spaces brings some advantages. These companies are already used to share and cooperate. They already have a mental setup, they know the advantages to cooperate and can use in long-distance cooperation the same intellectual instruments and procedures which they use on their short-distance projects.

There are already many international aggregations of shared spaces: as the COWO network that counts around one hundred coworking spaces in Italy or the Fab Lab network that counts approximately more than one thousand Fab Labs around the world.

The INCS is a small network where all the spaces know each other, host and send members to each other.

The Network is currently under the testing phase and the rules and characteristics of the network are not yet fixed. It is in a trial and error phase where some tools are under test and development.

It is a network where diversity is one important feature, diversity of spaces included. Not only multifactories can be part of the network but also fab labs, maker spaces or Cultural shared spaces are welcome, all shared spaces that have three characteristics: diversity, a community-based governance system and heterogeneity.

Senge (2006) says that "You cannot have a learning organization without shared vision. Without a pull toward some goal which people truly want to achieve, the forces in support of the status quo can be overwhelming." The INCS is a way to establish a highly structured value chain, organized in a non-capitalistic way, in order to produce complex goods and services.

The overall vision of this research is to build a system in which it is possible to change how things are produced and, in the long run, to shift from an economic paradigm to a new one. This should affect the whole economic system and it should be possible to be competitive with the existing one also in building complex products.

A strong and structured network is a way to accelerate partnerships between equals and the ability to create strong and large enough teams to reach an international market, anticipating traditional companies and avoiding the replication of the capitalistic model in relationships between shared workspaces.

New business ideas born inside self-regulated shared working spaces can be competitive only if they have the opportunity to enter the global market.

An independent network of physical spaces is a very valuable chance to accelerate the process of social change taking place in these open contexts.

Places where crowdworkers from other spaces don't just have the right to enter, to work, to make a meeting, but where they know people, stories, issues, where they can develop projects, ask for help, give a hand and feel at home.

The Structure of the International Network of Collaborative Spaces

The INCS has been conceived by the Authors a multi-level connectivity system.

The idea at the base of the Network is simple: to create a multi-dimensional communication and exchange system in order to give answer to common needs to share, to exchange, to meet.

The Network is made up by places which are different but share some characteristics:

- A – They are shared workspaces
- B – There is a focus on production of goods
- C – They are community projects

The network is built in order to facilitate exchanges: As a goal, every node in the network should not be further than 300 km from at least another node, but the network would grow with different speeds and logics in every Country, depending on local needs, the INCS is mainly based on interactions between people who work in the spaces.

The structure of the network doesn't have a top-down approach, this means for instance that there isn't a central office which manages relationships between members and every node acts in full autonomy.

To manage the network, at the beginning of 2018, a small group of people are involved in developing the system and working on agreements, exploitation, and so on, but this group of initiators is not intended to be a "board". It's just a starter group for the kick-off and to directly support the project while it is in the start-up phase, and it's planned to shift to an open community of developers, coming from the network itself, who will take care of it and let it grow.

So, the Network is designed in order to keep the "infrastructural" parts to a minimum. Basically, the only centralized environments are a website (www.multifactory.org) which explains what the network is, how it works and shows the different spaces with a brief description, a Facebook page and a YouTube Channel "Do It Together" (these two can be easily organized to have a shared management).

The Multifactory Network relies on a stable, strong network of relationships based on a Shared Governance System and a widespread management system. No one is a key factor and if a node goes away, others can replace it.

The Core of the Network: The Multi-Level Connectivity System

Most of the network is about "exchange", so the real backbone of the whole network is the Multi-level Connectivity System. The goal is to foster the exchange between people working and living in different cities and nations and to reach this goal.

The Multi-level Connectivity System within the INCS is made of four levels, that represent four steps to know each other:

First Level: Video-Tutorial Presentations

The first step to get to know each other is to make a presentation, and the simplest way to do this is to use a video. A video-tutorial-presentation is a short video focused on one person and divided into two parts: the first one is a video tutorial, the second one is a video presentation of the person who made the tutorial.

The first part is a tutorial, as sharing knowledge is a key point to make the network grow and it's a way to show the personal style of each person. The second part is a self-presentation: where they work, how they work, what they like, where they find inspiration, what they're working on.

The idea is to first give something to the community (with the tutorial) and then to present activity, and projects. All the videos are similar and follow the same format and can be found on a YouTube channel (do it together). The YouTube channel is organized with tags and data to help being found by everyone.

Second Level: The Electronic Window

The "electronic window" is a direct way to establish connections between people who work in different cities. In every space which is part of the network there is a window, that is an electronic device based on raspberry PI. Each window can be opened on another one, but just one connection can be established for each pair of windows. Nevertheless, it's possible to choose another space just by pushing a button.

Each e-window is open all day long from one space to another. If there are two persons close to the two windows, they can talk and interact.

E-windows are based on low-cost technologies (Raspberry Pi, a flat monitor, a webcam, Linux and a few other electronics) and are intended to be just e-windows: they don't have keyboard, mouse, browser, and other programs installed.

If people want to meet in front of the window, they act exactly as if they would plan to meet someone in a physical place.

Third Level: Meet People From Other Projects Inside the Network

Every place has a similar mood and there is an electronic window in it. This way all members belonging to a workspace that is part of the network have the opportunity to know something about people who work in other spaces, through video-tutorial-presentations and e-windows.

At this point, maybe some members could be interested in someone else's project, or just want to see in person what others are doing. In this phase, it's likely that there would not be established relationships yet. So, it's important to have places that are quite close to one another, as this makes it possible to visit them in one day. That's why in the network each node has to be not further away than 300 km from another node. On the other hand, key nodes in big cities are interesting as well, as big cities can be easily reached and it can happen that people might have to go there already for other reasons.

This makes it easy to go to another space and start exploring what others are doing, as it lowers the initial barriers.

Fourth Level: Making Projects Together, The Free Exchange Program

The fourth level is to establish long-term relationships between people working in different places and to create projects together. With the Free Exchange Program, every person working in one space within the network can go and work in another soace at no fees and no expenses, but this is subjected to some conditions. First, the need to clearly explain the project to the hosting workspace (which can refuse the request), second, the need to give something to the community which hosts: a workshop, a lesson, an object, something that improves the place that hosts or the skills of its members in order to give a contribution to the growth of the community.

These conditions apply because the Free Exchange Program is not intended to be a cheap way to have an office in many cities, but a tool to facilitate real exchanges within the network, and to lower anxiety and doubts about the opportunity to spend a period abroad. With the Free Exchange Program, members of every space know that they can go to another space and find people who are similar to them.

Modelling the Growth

During the search the various spaces were found mainly through word of mouth, or external reports, or internet search activities or social networks using keywords. Therefore, it can be said that the sample is composed of randomly chosen units, without a predefined search scheme. Once the data related to the dimensional class of the spaces visited during the research had been analyzed, a distribution was obtained as shown in Figure 1.

Although the sample is limited, analyzing the data it seems possible to assume that the distribution may follow the power laws distribution. The spaces also seem to follow some other typical rules of urban agglomerations, and when they can no longer expand in the spatial sense they increase the density of the population by reclaiming the spaces at the beginning left unstructured or extending vertically.

Future Research should better investigate these topics, trying to define a model of growth and checking to a larger scale if the distribution really follows the power laws.

Figure 1. Number of visited shared workspaces for each dimensional class

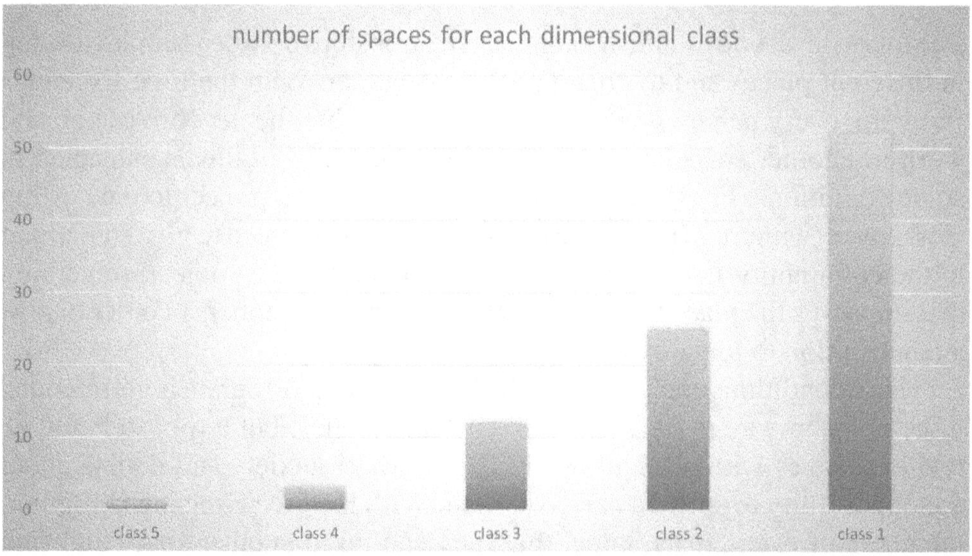

CONCLUSION

The experiences studied during the research come from different cultural backgrounds and have undergone different evolutionary paths. As many experiences are still pioneering and experimental, often the result of local private initiatives, the coordinators and members are strongly oriented to follow and develop the internal relational dynamics.

However, the research has shown that there is a consistent overlap between lifestyles, values, organizational models and functioning mechanisms. This suggests that a scalable system of professional relations can be hypothesized, reproducing the same collaborative dynamics present within each shared workspace between shared workspaces.

In this perspective, it is possible to imagine a supranational system of autonomous spaces but linked by value relationships and professional collaborations, able to constitute a chain of skills and capabilities able to imagine, design, develop and implement complex products and services. This kind of collaboration could take up and expand the same principles that underpin the invisible factory.

To make this vision possible, it is first of all necessary to base the collaborative structure on the same concepts that underpin the collaboration within the spaces, that is, on the mutual knowledge between members and on the sharing of experiences and spaces. This can be accomplished essentially on two levels, first in the virtual space of remote relationships, through tutorials and electronic window, then in real space, through travels and the free exchange program.

Once the electronic and organizational tools of the system have been fully developed, it will be possible to start to start concrete experimentations on the joint transnational development of complex products and services between shared workspaces.

If the spatial distribution of shared workspaces followed the law of power, one could imagine clustering spaces, for example by agglomerating them in an aggregate form around the most relevant dimensionally. If it is clear that small or medium-small spaces cannot think of making global products, it is also true that huge, large or medium-large spaces could have the strength, if inserted into a system of stable relations, to develop and convey on the market complex products and services, also made in partnership with smaller spaces, fully realizing on a large scale that revolution in the way of producing that each space is experimenting locally.

REFERENCES

Senge, P. M. (2006). The Fifth Discipline: The Art and Practice of the Learning Organization. Academic Press.

ADDITIONAL READING

Aigrain, P. (2014). *Sharing: Culture and the Economy in the Internet Age.* Amsterdam, The Netherlands: Amsterdam University Press.

Amidon, D. M., Formica, P., & Mercier-Laurent, E. (2005). *Readings on Knowledge Economics: Emerging Principles, Practices and Policies.* Tartu University Press.

Anderson, C. (2010). How web video powers global innovation. *TEDGlobal.* Retrieved June, 12, 2013 from http://www.ted.com/talks/chris_anderson_how_web_video_powers_global_innovation.html

Anderson, R. (1999). *Mid-course correction: toward a sustainable enterprise.* Chelsea, MA: Green publishing Company.

Anderson, R. (2009). The business logic of sustainability. *TEDX.* Retrieved July, 2013, from http://www.ted.com/talks/ray_anderson_on_the_business_logic_of_sustainability.html

Antagata, W., (2007). *Libro Bianco sulla creatività.* Roma: Commissione sulla Creatività e Produzione di Cultura in Italia/Ministero per i Beni e le Attività Culturali.

Ariely, D. (2012). What Make us feel Good About our Work. *TED talk. Rio de la Plata.* Retrieved July, 2013, 23, from http://www.ted.com/talks/dan_ariely_what_makes_us_feel_good_about_our_work.html

Baier, A., Muller, C., & Werner, K. (2013). *Stadt der Commonisten. Neue urbane Raume des Do it yourself* [City of the Commonists. New urban spaces of do it yourself.]. Bielefeld, Germany: Transcript. doi:10.14361/9783839423677

Bettoni, M. C. (n.d.). *The Essence of Knowledge Management: A Constructivist Approach.* Institute for Methods and Structures.

Bollier, D. (2008). *Viral spiral. How the Commoners Built a Digital republic of their Own.* New York, NY: The New Press. Retrieved from http://www.viralspiral.cc/sites/default/files/ViralSpiral.pdf

Brabham, D. C. (2013). *Crowdsourcing*. Cambridge, MA: MIT Press Essential Knowledge.

Broderick, D. (2001). *The spike: How Our Lives Are Being Transformed by rapidly Advancing Technologies*. New York, NY: Tom Doherty Associated.

Carr, N. (2007). The Ignorance of Crowds. *Strategy Business*. Retrieved November 15, 2014, from http://www.strategy-business.com/article/07204?pg=all

Catelli, G. (1976). *La società marginale. Contadini, Sottoproletariato ed emarginati come societa' negative.* Roma, Italy: Citta' Nuova.

Caves, R. E. (2000). *Creative Industries, Contracts between Art and Commerce*. Cambridge, MA: Harvard University Press.

Chapain, C., Comunian, R., & Clifton, R. (2011). Location, location, location: exploring the complex relationship between creative industries and place. *Creative Industries Journal, 1*(3), 5-10. Retrieved from http://www.ingentaconnect.com/content/intellect/cij/2010/00000003/00000001/art00002

Chesbrough, H. (2003). *Open Innovation: The New Imperative for Creating and Profiting from Technology*. Boston: Harvard Business School Press.

Chesbrough, H. W., & Di Minin, A. (2008). *Open: Modelli di business per l'innovazione*. Milano, Italy: Egea.

Christopher A., & Dominique C. (2007). The strength of weak cooperation: an attempt to understand the meaning of web 2.0. *International Journal of Digital Economics, 65*, 51-65.

Communication from the Commission to the European Parliament, the Council, the European economic and social Committee and the Committee of the regions. (2011). *A renewed EU strategy 2011-14 for Corporate Social Responsibility* {COM(2011) 681 final}. Brussels: EU. Retrieved from http://eur- lex.europa.eu/LexUriServ/LexUriServ.do?uri=COM:2011:0681:FIN:EN:PDF

Communication from the Commission to the European Parliament, the Council, the European economic and social Committee and the Committee of the regions. Social Business Initiative. (2011). *Creating a favorable climate for social enterprises, key stakeholders in the social economy and innovation* {SEC(2011) 1278 final}. Brussels: EU. Retrieved from http://eur- lex.europa.eu/LexUriServ/LexUriServ.do?uri=COM:2011:0682:FIN:EN:PDF

Communication from the Commission to the European Parliament. the Council, the European economic and social Committee and the Committee of the regions. (2011). *An Agenda for new skills and jobs: A European contribution towards full employment* {26.11.2010 COM(2010) 682 final/2}. Strasbourg, France: EU. Retrieved from http://eur- lex.europa.eu/LexUriServ/LexUriServ.do?uri=COM:2010:0682:REV1:EN:PDF

Coy, P. (2000, August 28). The creative economy. Which companies will thrive in the coming years? Those that value ideas above all else. *Business Week Magazine*.

Creative Metropoles. Situation Analysis of 11 Cities: Final Report. (2010). Retrieved from Creative Metropoles Interreg IVC Program web site: www.creativemetropoles.eu

Cressey, P., & Kelleher, M. (2003). The Conundrum of the Learning Organisation - Instrumental and Emancipatory Theories of Learning. In B. Nyhan, P. Cressey, M. Kelleher, & R. Poell (Eds.), Learning Organisations: European perspectives, theories and practices. Luxembourg: CEDEFOP.

Deskmag, C. E. (2017). *2017 Global Coworking Survey*. Retrieved from http://www.deskmag.com/en/background-of-the-2017-global-coworking-survey

Drew, J., Sundsted, T., & Bacigalupo, T. (2009). *I'm Outta Here: How Coworking is Making the Office Obsolete*. Austin, TX: NotanMBA Press.

Drucker, P. (1993). *Post-Capitalist Society*. Oxford, UK: Butterworth Heinemann.

Eck, J. (2005). Struggling with the creative class. *International Journal of Urban and Regional Research*, *29*(4), 740–770. doi:10.1111/j.1468-2427.2005.00620.x

European Commission. (2010a). *Green Paper. Unlocking the potential of cultural and creative industries*. Brussels: EU.

European Commission. (2010b). Europe 2020. A European strategy for smart, sustainable and inclusive growth. Brussels: EU.

Frank, M., Roerigh, P., & Pring, B. (2017). *What To Do When Machines Do Everything: How to Get Ahead in a World of AI, Algorithms, Bots, and Big Data*. Cambridge, UK: Polity Books.

Gansky, L. (2012). *The Mesh: Why the Future of Business Is Sharing*. New York: Portfolio Trade, Penguin Group.

Genevieve, V., & DeGuzman, A. I. (2011). Working in the UnOffice: A Guide to Coworking for Indie Workers, Small Businesses, and Nonprofits. San Francisco, CA: Night Owls Press.

GNP. (2013). *Tem(p)i e luoghi del lavoro flessibile.* Retrieved from http://www.giornatanazionaledellaprevidenza.it/terza-giornata-gnp2013

Gordon, R., & Brynjolfsson, E. (2013). The future of work and innovation debate. *Ted2013*. Retrieved July 15, from http://blog.ted.com/2013/04/23/the-future-of-work-and-innovation-robert-gordo- and-erik-brynjolfsson-debate-at-ted2013/

Harley, C. K. (2013). *Slavery, the British Atlantic Economy and the Industrial Revolution*. Oxford, UK: Oxford university Press.

Howkins, J. (2001). *The Creative Economy: How People Make Money From Ideas*. New York: Penguin Group.

Howkins, J. (2005, September). Enhancing creativity. In Creative Industries. A symposium on culture based development strategies. New Delhi: Malvika Singh.

Huston, J. L. (2003). *Calculating the Value of the Union: Slavery, Property Rights, and the Economic Origins of the Civil War.* Chapell Hill, NC: The University of North Carolina Press.

Jackson, T. (2005). *Motivating Sustainable Consumption: A Review of Evidence on Consumer Behavior and Behavioral Change.* Paper published by the Centre for Environmental Strategy, University of Surrey. Retrieved from www.epa.gov/sustainability/workshop0505/5d_Jackson_Tim.pdf

Jacobs, J. (1961). *The death and life of great American Cities*. New York: The Random House.

Kurki, L., & Manoliu, M. (2011). Opinion of the European Economic and Social Committee on Innovative workplaces as a source of productivity and quality jobs (own-initiative opinion). *Official Journal of the European Union*. Retrieved from http://eur-lex.europa.eu/LexUriServ/LexUriServ.do?uri=OJ:C:2011:132:0022:0025:EN:PDF

Micelli, S. (2011). *Futuro artigiano*. Venezia, Italy: Marsilio Editore.

Miroslav, B., Maselli, I., & Martellucci, E. (2012). *Workplace Innovation and Technological Change*. Brussels: Centre for European Policy Studies Special Reports. Available at SSRN: http://ssrn.com/abstract=2147619

Morace, F. (2011). I paradigmi del futuro. Lo scenario dei trends [The paradigms of the future. The trends scenario]. Busto Arsizio, Italy: Nomos Edizioni.

Namier, L. B. (1952). *Avenues of history*. London, UK: H. Hamilton.

Nordström, K. A., & Ridderstrale, J. (2007). *Funky Business Forever: How to enjoy capitalism*. Pearson Education.

Pot, F., & Associates. (2011 September). *Social innovation of work and employment, Challenge Social Innovation*. Paper presented at the workshop: Social innovation at work, Wien, Austria.

Raymond, E. S. (2001). *The Cathedral and the Bazaar: Musings on Linux and Open Source by an Accidental Revolutionary.* Sebastopol, CA: US O'Reilly Media.

Rivas, M., (2011). *From creative industries to the creative place. refreshing the local development agenda in small and medium size towns.* URBACT creative clusters project final report, Óbidos, PT.

Rombie, D. (2010). *The Entrepreneurial Dimension of the Cultural and Creative Industries. A report commissioned by the European Commission.* Utrecht School of the Arts.

Rushkoff, D. (2011). *Life Inc: How Corporatism Conquered the World, and How We Can Take It Back.* New York: Penguin Random House Trade Paperbacks.

Sapelli, G. (1994). *Economia, tecnologia e direzione d'impresa in Italia.* Torino, Italy: Einaudi.

Sapelli, G. (1996). Responsabilità d'impresa. In *Tra mercato e nuova sovranità politica*. Milano, Italy: Guerini e Associati.

Shirky, C. (2008). *Here Comes Everybody: The Power of Organizing Without Organizations*. New York, US: Penguin Group.

Sioli, M. (2004). *Metropoli e natura sulle frontiere americane: Dalle non-città indiane alla città di Thoreau, dalle metropoli industriali alla città ecologica* [Metropolis and nature on the American frontier: From Indian non-cities to Thoreau's city, from industrial metropolises to the ecological city]. Milano, Italy: Franco Angeli.

Srnicek, N., & Williams, A. (2015). *Inventing the Future: Postcapitalism and a World Without Work*. London, UK: Verso Books.

Tapscott, D., & Williams, A. D. (2007). *Wikinomics: How Mass Collaboration Changes Everything*. New York: Portfolio Penguin Group.

Towse, R. (2002). Book Review of Creative Industries. *Journal of Political Economy*, 234–237. doi:10.1086/324388

Van Heur, B. (2010). *Creative Networks and the City: towards a Cultural Political Economy of Aesthetic Production*. Bielefeld, Germany: Transcript. doi:10.14361/transcript.9783839413746

Verwijnen, J. (1999). The Creative City's New Field Condition. Can Urban Innovation and Creativity Overcome Bureaucracy and Technocracy? In J. Verwijnen & P. Lehtovuori (Eds.), *Creative Cities. Cultural Industries e Urban Development and the Information Society*. Helsinki: UIAH Publications.

Williams, S. (2012). Free as in Freedom. In *Richard Stallman's Crusade for Free Software*. Sebastopol, CA: O'Reilly Media.

About the Authors

Giulio Focardi has a master degree in Economic History and his main topics of interest as a researcher and professional are related to Community Economy, Collaborative Economy and Social Economy. He used to work as a consultant, mostly in the HR field and strategic development and multidimensional planning. He developed various mathematical models of social systems and is co-author of several books. His working method starts from the systemic analysis of situations, developing intervention models through a trace-back process, from the phenomenon to the causes. He is the Co-Founder and CEO of Osun WES, a consulting firm specialized in developing companies as collaborative systems, and he is one of the initiators and president of the board of R84 Multifactory Mantova.

Lorenza Victoria Salati has a master degree in Political Sciences (with focus on economic history and visual antropology) and a specialization at the School of Cinema and New Media of Milan. She developed methods derived by visual anthropology as a support to the self-perception of singles as change makers. She used to be a documentary filmmaker to explore agricultural and ethnical issues in Africa (Burkina Faso, Mozambique, Senegal). In the meanwhile, facing up to new communication needs emerged from the opportunities offered by the Web 3.0 she investigated new forms of expression and innovative methods of storytelling. She made many documentaries and is co-author of several books. She is the co-founder of the Italian firm Osun WES and is the co-owner of a German based company specialized in technological tools to support collaborative projects. She is one of the initiators and vice-president of R84 Multifactory Mantova.

Index

T

V

W

Y

www.ingramcontent.com/pod-product-compliance
Lightning Source LLC
La Vergne TN
LVHW061534290926
849162LV00004B/790
* 9 7 8 1 5 2 2 5 7 0 8 9 9 *